Looking for Longitude

A Cultural History

Eighteenth-Century Worlds

Series Editors:
Professor Eve Rosenhaft (Liverpool) and Mark Towsey (Liverpool)

Eighteenth-Century Worlds promotes innovative new research in the political, social, economic, intellectual and cultural life of the 'long' eighteenth century (c. 1650–c. 1850), from a variety of historical, theoretical and critical perspectives. Monographs published in the series adopt international, comparative and/or interdisciplinary approaches to the global eighteenth century, in volumes that make the results of specialist research accessible to an informed, but not discipline-specific, audience.

Looking for Longitude

A Cultural History

Katy Barrett

Liverpool University Press

First published 2022 by
Liverpool University Press
4 Cambridge Street
Liverpool
L69 7ZU

British Library Cataloguing-in-Publication data
A British Library CIP record is available

ISBN 978-1-80207-053-8

Typeset by Carnegie Book Production, Lancaster

For Polly and Phil,
who first taught me how to take a line for a walk,

and for Nick,
who prefers a Hogarthian wanton line.

Contents

Figures

Acknowledgements

LIKE THE EIGHTEENTH-CENTURY SEARCH for the longitude, this book would not have reached a conclusion without the input of many, many people. My first thanks must go to the other members of the 'Board of Longitude' project team for their encouragement and ideas: to Rebekah Higgitt, Alexi Baker, Nicky Reeves, Eóin Phillips, and especially to Sophie Waring. Likewise, I owe thanks to a number of academics and colleagues who have been constant sources of advice and support over the years: to Faramerz Dabhoiwala, John Robertson, Alison Petch, Katie Eagleton, Kim Sloan, Kate Retford, Melanie Vandenbrouck, Martha Fleming, Ludmilla Jordanova, and Mark Hallett. As Joshua Reynolds said of his literary friends, they have 'formed my mind … and brushed off from it a deal of rubbish'.

More specifically, Part I benefited from the input of Jim Bennett, David Bryden, Gillian Hutchinson, and Sheila O'Connell. Part II is better thanks to Greg Lynall and Pat Rogers. For comments on Part III, I am grateful to Faramerz Dabhoiwala, James Delbourgo, and Patricia Fara. I have, likewise, benefited from more general discussions with Charlotte Ashby, Jonathan Betts, James Davey, Jean-Francois Gauvin, Robert Hicks, Laura Humphreys, Richard Johns, Stephen Johnston, Matthew Jones, Larry Klein, Sarah Longair, Joserra Ramon Marcaida, Alexander Marr, Rory McEvoy, Laurel Peterson, Pamela Smith, Clifford Siskin, and Rebecca Wallis. The ideas for many parts of this book were honed and developed with the input of colleagues and delegates at many conferences, workshops, and seminars over the years. Without the support, keen eyes, and perceptive comments of Caitlin Wylie, Geoff Belknap, and most especially Gregory Lim, this work would have been considerably less polished. Renewed thanks to all of the friends and colleagues who have put me up, or simply put up with me, over the years taken to bring this book to fruition.

Particular thanks must go to Ally Lee and the team at Liverpool University Press for bringing this research to book form, and to the anonymous peer reviewers who saw the merit of the project and advised on its development.

I have been assisted with great patience and kindness by librarians and curators at a range of different collections: the Cambridge University Library; the National Maritime Museum; Gonville and Caius, St John's, Trinity, and Newnham colleges, Cambridge; the Whipple Library and Fitzwilliam Museum Print Room, Cambridge; Magdalen College, Oxford; the Bodleian Library, Oxford; the British Museum Print Room; the V&A Department of Sculpture, Metalwork, Ceramics and Glass; the British Library; the National Library of Scotland; the Royal Society Archives; Special Collections at University College London; the Guildhall Library; the Royal Geographical Society; the Science Museum; the Conservation Department at Tate Britain; Sir John Soane's Museum; the Wellcome Library; the Bibliothèque Nationale de France; the Elmer L. Andersen Library, Minneapolis; the Houghton Library, Harvard University; the New York Public Library; the Beinecke Rare Book and Manuscript Library, Center for British Art, Medical Library, and Lewis Walpole Library at Yale University; Denver Public Library; Ohio University Libraries; and the William L. Clements Library, Michigan. Likewise, this book would have been impossible without Early English Books Online (EEBO), Eighteenth-Century Collections Online (ECCO), the British Museum collections online, the digital collections of the Yale Center for British Art and Lewis Walpole Library, and the Burney Newspaper Collection. The Paul Mellon Centre made it possible for me to carry out five-weeks' research in the USA.

There are five people in particular without whom this book and I would not be where we are today. I am endlessly grateful to Simon Schaffer and Richard Dunn for their time, patience, and expertise as supervisors, colleagues, champions, and friends; to my parents Phil and Polly Barrett for their support, encouragement, and wisdom over so many years; and to Nick Beever for being there to get me through each and every day.

Abbreviations

Analysis	William Hogarth, *The Analysis of Beauty* (ed. Ronald Paulson) (Yale, 1997).
HJ	'Harrison Journal' MS, Private Collection.
ODNB	*The Oxford Dictionary of National Biography* online, http://www.oxforddnb.com.
Principles	John Harrison, *The Principles of Mr. Harrison's Time-Keeper, with plates of the same. Published by order of the Commissioners of Longitude* (London, 1767).
RGO 14	Papers of the Board of Longitude, Cambridge University Digital Library, RGO 14/1–68, http://cudl.lib.cam.ac.uk/collections/longitude.

Introduction
What Was the Problem with Longitude?

I don't understand your Longitude, I never heard such a cramp Word in my Born-days.[1]

THERE IS ONLY ONE KNOWN IMAGE of someone attempting to solve the longitude problem in the eighteenth century. William Hogarth's (1697–1764) final image in the series *A Rake's Progress* of 1735 shows a madman scribbling a series of longitude diagrams on the wall of Bedlam, the London madhouse (Figure 1). This book tells the story of those diagrams and why a problem like finding longitude at sea was of sufficient cultural importance to appear in Hogarth's image.

A Rake's Progress is the tale in eight parts of a young man, Tom Rakewell, corrupted by London society. Starting out as the naïve heir to his miserly father's fortune, Tom moves to London to become a fashionable gentleman, holds an expensive levée, visits a brothel, and narrowly avoids being arrested for debt. He attempts to resurrect his declining fortune by marrying a rich widow, but soon loses his new riches in a gambling den and finds himself in Newgate debtors' prison. Hogarth shows Tom ending his days distraught in Bethlehem Hospital. He has become disorientated in his attempts to navigate London society.

Behind him, a more shadowy figure is trying to solve the longitude problem on the wall of his cell. His map of the world, his series of diagrams, and his simple label 'longitude' denote a problem in navigational science that

1 Saavedra, *The history of the renowned Don Quixote*, iii, p.262.

Figure 1: William Hogarth, *A Rake's Progress*, Plate 8 (1735); engraving, 354 x 405mm, Harris Brisbane Dick Fund, 1932, Metropolitan Museum of Art, New York.

Figure 1: detail.

obsessed Hogarth's contemporaries but is now little known: the problem of finding longitude at sea. This longitude lunatic therefore gives us a route into the wealth of sources that commented on longitude, in word and image, both seriously and satirically, throughout the eighteenth century, but are now forgotten. In the following chapters we trace how 'the longitude' articulated a series of broader problems—visual, mental, and social—for contemporary society.[2] The cultural ideas, practices, and disagreements that shaped the longitude problem also shaped the society which Hogarth so vividly depicts destroying Tom Rakewell.[3] Finding longitude—orientating navigators at sea—might also provide a means of orientating society.

2 Hardy, *Theory of the Moon*, p.9.

3 This book thus attempts, in part, to meet the challenge posed by Jordanova in *The Look of the Past.*

'the GRAND PROBLEM, to determine the Longitude of our Ship at Sea'

By the early eighteenth century, European ships had been ploughing the oceans for centuries, but the problem of accurate navigation was becoming increasingly pressing as colonial trade routes were cemented and Britain entered a period of near-continuous war with France. Since 1567, a number of European powers had offered prizes for successful methods of finding longitude at sea, including Spain, France, and the Dutch Republic.[4] Latitude is fairly easy to ascertain from celestial observations as it is linked to the points of the poles and the line of the equator, but longitude has no such fixed points and thus presents problems for navigators. In 1714, the British government therefore passed the Longitude Act, which initiated the publications, images, and debates examined in this book. The Act outlined how 'nothing is so much wanted and desired at Sea, as the Discovery of the Longitude, for the Safety and Quickness of Voyages, the Preservation of ships, and the Lives of Men'.[5] It created the possibility of funding to support the development and trial of proposals up to the sum of two thousand pounds, and offered rewards of ten, fifteen, or twenty thousand pounds for finding longitude at sea to varying degrees of accuracy (to within sixty, forty, and thirty geographical miles, respectively). Half would be paid for improving the safety of ships close to the shore, and the other half after a successful trial to the West Indies. The Act also established a group of commissioners to judge proposals, organise trials, and decide on the final award. The Commissioners were composed of the great and the good of contemporary natural philosophy, politics, and seafaring. They included, *ex officio*, the Speaker of the House of Commons, the Lord High Admiral and admirals of the Red, White and Blue squadrons, the Astronomer Royal, the President of the Royal Society, and professors from Oxford and Cambridge, as well as nine named lords and MPs. Crucially, they were to award money when proposed methods had been 'tried and found practicable and useful at Sea'. This wording would prove particularly problematic in the ensuing decades.

The Act outlined from the start how 'several Methods have already been discovered, true in Theory, though very difficult in Practice', making clear that potential solutions were known in 1714 but needed development. It was Sir Isaac Newton (1643–1727) who outlined the solutions then credited as

---

4 Andrewes, *Quest for Longitude*, p.2; Dunn and Higgitt, *Finding Longitude*, pp.12–33; Baker et al., *The Board of Longitude*.

5 RGO 14/1: 11r.

possible to the House of Commons during the passage of the Act, as well as the limitations to each:

> One is by a Watch to keep Time exactly. But, by reason of the motion of a ship, the variation of heat & cold, & the difference of gravity in different Latitudes, such a Watch hath not yet been made.
>
> Another is by the Eclipses of Jupiter's Satellites. But by reason of the length of Telescopes requisite to observe them & the motion of a ship at sea, those Eclipses cannot yet be there observed.
>
> A third is by the place of the Moon. But her Theory is not yet exact enough for this purpose. It is exact enough to determine her Longitude within two or three degrees, but not within a degree.
>
> A Fourth is Mr Ditton's project. And this is rather for keeping an account of the Longitude at sea than for finding it if at any time it should be lost, as it may easily be in cloudy weather.[6]

The first three methods were repeatedly reformulated in different proposals, along with a further suggestion for mapping patterns of magnetic variation or dip. All required instruments to be perfected for use at sea, as well as the simplification of complex mathematical calculations. The fourth method was a specific proposal by William Whiston (1667–1752) and Humphrey Ditton (1675–1715), which had helped to instigate the passage of the Act and proposed finding longitude by firing mortars at agreed times from ships moored at regular intervals across the oceans.[7] Vessels would then calculate their longitude from the difference of time it took the sound and light from the mortar to reach them. It is this, then infamous, proposal that Hogarth's longitude lunatic references with the mortar and projectile that he draws on the wall of Bedlam.

After the passage of the 1714 Act, the promise of reward money and fame created a flood of pamphlet proposals, but no proposal was taken seriously by the Commissioners of Longitude until the 1730s. In 1727–28, the Yorkshire clockmaker John Harrison (1693–1776) visited London with plans for his first timekeeper (now known as H1), which he discussed with the Astronomer Royal Edmond Halley (1656–1742) and fellow clockmaker George Graham (1673–1751).[8] They were suitably impressed for it to be trialled on a voyage to Lisbon in 1736 on the *Centurion*, and the timekeeper's success was responsible for the first recorded meeting of the Commissioners of Longitude in 1737.

6 Newton, 'Report', Newton Papers, MS Add.3972.

7 Whiston, *A new method*.

8 John Harrison, H1 (1735); timekeeper, height 673mm, Royal Museums Greenwich.

Over the next thirty-six years, the Commissioners funded Harrison to create three further timekeepers (H2, H3, and H4), and negotiated, increasingly acrimoniously, with him over what these timepieces needed to do to be proved 'practicable and useful at sea', as required by the act of 1714.[9] Over the course of two trials to the West Indies, to Jamaica in 1761–62 and Barbados in 1764, and four further parliamentary acts, the Commissioners struggled to establish what Harrison needed to do to merit the 'great reward'. Harrison is thus one of the key protagonists in our story, whose life was fundamentally changed by the longitude problem and the web of discussions surrounding it.

The famous reward money, the constant flood of pamphlet proposals, and the Board's increasingly public negotiations with Harrison brought longitude consistently to public attention over the eighteenth century. The problem was discussed and referenced in every conceivable type of contemporary literature, from plays to poems, religious tracts to legal discussions, and a host of visual material beyond Hogarth's print. Longitude became a joke, a throwaway remark, and an innuendo.[10] Attending to this group of references opens to us the array of discussions that took place outside of the Board, and the naval or natural philosophical sphere, and allows us to see how Harrison's engagement with the Commissioners was shaped by this broader context.[11] Both in direct proposals for solving the problem and in wider commentary on society, the

---

9 Eight further longitude acts passed through Parliament between 1714 (12 Anne c.15) and the 1774 Act (14 Geo.III c.66) that repealed and reformulated it. Acts in 1740 (14 Geo.II c.39), 1753 (26 Geo.II c.25), 1762 (2 Geo. III c.18), 1765 (5 Geo. III c.11), and 1770 (10 Geo. III c.34) gave the Commissioners further money, mostly to fund Harrison. 1740 made this money available for coastal surveys, and 1753 added five new commissioners. Act 3 Geo. III c.14 in 1763 was an attempt to award Harrison £10,000 in two instalments in return for disclosing the principles of his watch, while a second act in 1765 (5 Geo. III c.20) made this award on similar conditions, along with awards to Leonard Euler and Tobias Mayer for work on the lunar distance method. An Act in 1773 (13 Geo. III c.77) voted £8,750 to Harrison as a parliamentary gratuity, the closest he came to receiving the reward.
John Harrison, H2 (1739), timekeeper, height 686mm; H3 (1757), timekeeper, height 622mm; H4 (1759), timekeeper, 165 x 124 x 28mm, all Royal Museums Greenwich.

10 This was not new in 1714—*Curious Enquiries* published in 1688, for example, presented a satirical proposal for finding longitude using 'powder of sympathy'—but the Longitude Act increased the volume and frequency of satire.

11 Some few of the satirical publications on longitude are included by Bouvier in *English Books*. Indeed, Owen Gingerich has made a joke of the more unusual published proposals. See 'Cranks and Opportunists' in Andrewes (ed.), *Quest*, pp.133–48. This volume presents the major academic discussion of the longitude problem to date. The conference from which it stems, likewise, inspired Dava Sobel's successful popular work *Longitude*. Elsewhere, both Lynall and Freiburg have considered wider discussions of longitude in relation to satire, both drawing on the arguments of this

following chapters examine how the word 'longitude' became a focus for public discussion, a cultural 'keyword' that represented, explained, and suggested a solution to a set of wider social concerns.[12] This book highlights this keyword by exploring a range of very different but significant places, sources, and people. By looking at the range of discussions that took place in London, in the plethora of printed materials that exploded into the eighteenth-century public arena, we can locate a network of practices, images, and ideas that unite around the problem of longitude. For the longitude problem had to be solved in London before it could ever be tried and tested at sea.

'A great variety of papers and Pamphlets'

Eighteenth-century London was a city brimming with papers and pamphlets. The lapse of the Licensing Act in 1695 had brought an end to all pre-publication censorship, allowing the proliferation of a range of print materials. The first daily newspapers, the *Post Boy* and *Post Man*, appeared in 1701 and 1702, and the periodical was born with the *Tatler* and *Spectator* in 1709 and 1711. Newspapers became a vehicle for extensive advertising, announcing publications, entertainments, new commodities, and even lost property.[13] Print became part of daily life: from newspapers, advertisements, and almanacs, to engravings, playing cards, handbills, and catalogues, this outpouring of printed ephemera became known by the London location from which it mostly emanated: 'Grub Street'.[14] In the following chapters we examine and sample a wide variety of this outpouring of material, highlighting the sheer weight of disparate sources that engaged with the longitude problem. Understanding the search for longitude requires considering a cast of characters from among the printers, publishers, and booksellers as much as politicians, navigators, or natural philosophers. The emphasis of this book is on print which was disseminated to the widest audiences, but manuscript sources also connect into these discussions. It would be virtually impossible to find this breadth of mentions of longitude across unexpected and obscure printed texts without using the online collections databases that are now

book, then in unpublished thesis form. See Lynall, 'Scriblerian Projections', pp.1–18, and Freiburg, 'Our never failing guide', pp.33–62.

12 See Williams, *Keywords*; for discussion of an idea becoming a 'set of representations' rather than a 'real life practice' see Stallybrass and White, *Politics and Poetics*, p.103.

13 Wigelsworth has discussed natural philosophical advertising among these developments in 'Bipartisan Politics', pp.517–40, and 'Navigation and Newsprint', pp.351–76.

14 Rogers, *Literature and Popular Culture*, *Eighteenth Century Encounters*, and *Grub Street*.

Figure 2: Anon., *The Coffee-house Politicians* (1733); engraving, 167 x 109mm, © Trustees of the British Museum.

available to historians.[15] Thousands of references to 'longitude' in the period surface from such a search, allowing for broad themes to emerge in how the idea of longitude was employed and critiqued. The following chapters look at both the most exemplary of such uses and those made by known commentators. In some cases, a swift succession of examples makes visible the number of similar discussions behind each reference. The many pamphlets specifically proposing a solution to the longitude problem can thus be looked at alongside more canonical eighteenth-century texts and in the light of much broader discussions.

The physical space where such printed materials were most widely discussed was the coffee house: important new spaces for discussion of contemporary

15 Eighteenth-Century Collections Online (ECCO) and Early English Books Online (EEBO).

ideas and debates (Figure 2).[16] London boasted more than two thousand such establishments by 1700.[17] Early longitude proposals were themselves a product of this print and coffee-house culture; part of widening public interest in science, as mercantile and commercial interest in navigation, especially longitude, helped to spread the discussion of longitude outside of traditional spaces of science, like the Royal Society.[18] One correspondent to the *Weekly Journal* in 1723 discussed how, on reading contradictory longitude pamphlet proposals, 'my Curiosity led me to *Randal's Coffee-house*'.[19] Pamphlets in particular became a print phenomenon. Starting at 6d. and ranging from one folded sheet to a stitched many-page publication, pamphlets offered a means to enter print quickly and cheaply with new and topical ideas.[20] They offered the perfect vehicle for longitude proposals to be published, and thus enter the public arena. Such pamphlets inundated John Flamsteed (1646–1719) when, as the first Astronomer Royal, he became a Commissioner of Longitude under the 1714 Act. He discussed in letters to his friend and previous assistant, Abraham Sharp (1653–1742), the 'great variety of papers and Pamphlets' with which he, and the London print market, were engulfed.[21] More than ninety longitude pamphlets had been published by the time Hogarth's *A Rake's Progress* appeared in 1735.[22] Below we will see how extensively longitude pamphlets engaged with each other and with discussions in wider media, as well as how broader discussions responded to pamphlet proposals. It was common practice in the period for library and bookshop contents to be sold at auction, with an extensive catalogue published in advance. It is striking how many such catalogues surviving from the eighteenth century contain at least one pamphlet on longitude.[23]

Printed images formed a distinct part both of this wide development of print material and of the broader eighteenth-century development of artistic forms from the novel to the landscape garden.[24] Vic Gatrell (b.1941) has

16 Suarez and Turner (eds.), *Cambridge History of the Book*, pp.20–3.

17 Stallybrass and White, *Politics and Poetics*, p.95.

18 Stewart, *Rise of Public Science*, especially 'The Longitudinarians' and 'Other Centres of Calculation'.

19 *Weekly Journal* (15 June 1723), p.2,578.

20 Raymond in *Pamphlets and Pamphleteering* describes pamphlets as 'the most public print medium' by 1700, p.26.

21 Flamsteed, *Correspondence*, Letter 1,368, p.715.

22 Bouvier, *English Books*, S-1.

23 These range, for example, from Ballard, *Bibliotheca Salmoneana*, to White, *Catalogue of the library*.

24 Ivins, *Prints and Visual Communication*, pp.1–2. Ogée has argued that serialisation of time in Hogarth's *Progresses* should be seen in the context of growing contemporary clock ownership. See 'Aesthetics and empiricism' in Ogée, *Dumb Show*, pp.182–5.

argued for prints as showing 'what was spoken among men rather than written in text or confessed in the salon',[25] and many historians have looked for clues to these discussions in the rich works of Hogarth, where we likewise found our rare image of longitude.[26] The price of prints ranged from 6d. to 10s., allowing for a spectrum of ownership from the apprentice to the wealthy.[27] Libraries would lend folios for an evening's viewing, and single prints were pasted onto the walls of fashionably decorated rooms, or pinned on street corners, in shops and inns. As the century progressed, 'print-shop windows were the picture galleries of the public'.[28] In 1735, Hogarth's own studio acted as such a gallery, where he hung the *Rake's Progress* paintings for visitors to view, and where they were memorised by hack engravers to copy later.[29] The ways in which the narratives and settings were (mis) remembered by these engravers is striking, showing their familiarity with the texts, settings, and contemporary characters that Hogarth referenced.[30] Our longitude lunatic was clearly recognised for his iconography and transferred to the piracies (Figure 3), but with a number of alterations that help us to think in the following chapters about the public discussions on which these

25 Gatrell, *City of Laughter*, p.12.

26 Ronald Paulson and Peter Wagner have examined Hogarth's entire *oeuvre* for iconographical significance and historical references. See Paulson, *Hogarth's Graphic Works* and *Hogarth: His Life, Art and Times*, and Wagner, *Reading Iconotexts*: 'Eroticism in Graphic Art' and 'Hogarth's Graphic Palimpsests'. Jenny Uglow, David Bindman, and Mark Hallett have used his life and works as a mirror of broader eighteenth-century London life: Uglow, *Hogarth*; Bindman, *Hogarth and his Times*; Hallett, *Spectacle of Difference* and *Hogarth*. David Solkin's *Painting for Money* has used Hogarth as a means to consider the commercialisation of eighteenth-century art and the history of consumption, and David Dabydeen has looked particularly for Hogarth's representation of 'blacks' and the setting of Walpole's commercial Britain: *Hogarth's Blacks* and *Hogarth, Walpole and Commercial Britain*. Bernd Krysmanski and Elizabeth Soulier-Détis have looked for hidden systems of iconography in Hogarth's works: for extreme religious iconoclasm and social transgression, or for a web of hidden Masonic imagery: Krysmanski, *Hogarth's Hidden Parts*; Soulier-Détis, *'Guess at the Rest'*. Wide-ranging edited collections from Bernadette Fort and Angela Rosenthal, Bindman, Frédéric Ogée, and Wagner have looked at Hogarth's response to questions of gender, natural philosophy, and social self-representation: Fort and Rosenthal (eds.), *Other Hogarth*; Bindman, Ogée, and Wagner, *Hogarth: Representing Nature's Machines*; Ogée, *Dumb Show*.

27 Hume uses Hogarth's increasing sales to discuss how the print market expanded alongside public thirst for art, in Hume, 'Economics of Culture in London', pp.529–30.

28 George, *Hogarth to Cruikshank*, p.17.

29 Nichols, *Biographical Anecdotes of William Hogarth*; and *Catalogue of his Works* comments that Walpole observed the pirate prints were made by Boitard, p.91.

30 Dabhoiwala has argued that hack versions of the *Harlot's Progress* spread Hogarth's message in Dabhoiwala, 'The Appropriation of Hogarth's Progresses', pp.1–19.

Figure 3: After William Hogarth, *[A Rake's Progress]* *[Plate 8]* (about 1735); etching, 353 x 435mm, courtesy of the Lewis Walpole Library, Yale University.

engravers drew. Longitude was not merely a topical reference in mathematics and natural philosophy, but a keyword in a vast literature, both serious and funny, a comment on London society as well as on navigation. Crucially, every one of these authors was sure that his or her readers would understand a reference to 'the longitude' as well as why it was problematic.

'our Swarme of hopefull Authors'

Yet, it is striking that we know more about most of Hogarth's engraved characters than we do about many of the authors of longitude proposals. Figures like Whiston, Flamsteed, and Harrison appear in a range of broader discussions, but many characters in the following chapters may be known only by the information on their work's title page. Flamsteed wrote to Sharp in 1714 about the 'Swarme of hopefull authors' that approached him.[31] These authors, who figure largely in this book, include those who appear nowhere else in contemporary literature, sometimes come with only titbits of information, yet often seem well aware both of others' publications and of wider discussions. Pamphleteers contributed to the longitude debate from all over the British Isles, Europe, and the American colonies. While this book largely focuses on discussions in London, the wealth of locations from which proposals arrived shows how a wider story of longitude in regional and continental context is yet to be written.[32] Among the characters who appear in Chapters 1 and 2, William Blundel's (1647–1723) works were published in Coventry, whereas James Clarke (fl. 1714–15) wrote from Wiltshire.[33] Dorotheo Alimari (1650–1727) sent his proposal from Venice, whereas Leonard Sturm (1669–1719) (who takes a turn in Chapter 5) published his in Nuremberg, sending it both to the British Commissioners of Longitude and to the Académie des Sciences in Paris.[34] Many pamphlets were addressed or dedicated to the Commissioners of Longitude, often by name, showing high awareness of the awards and judges established by the 1714 Act. Authors combined different combinations of print media and personal approach. Flamsteed told Sharp how in one month of 1714 he received a visit from 'a couple of young Non conformist preachers from Worksop in … Derbyshire', a printed pamphlet from the clockmaker William Hobbs (1658–fl.1716) (also in Chapter 2), another from Case Billingsley (fl.1714), and a letter from

31 Flamsteed, *Correspondence*, Letter 1,366, p.712.

32 On German longitude pamphlets see Köberer, 'German contributions'.

33 Blundel, *To all and Singular*; Thacker, *Longitudes examin'd*; Clarke, *Essay*.

34 Alimari, *New method*; Sturm, *Project*.

Conyers Purshall (fl.1705–7).[35] Each of these published their proposals,[36] but Purshall and Billingsley also produced short handbills advertising their schemes,[37] which we can imagine pinned to coffee-house or inn walls, while their pamphlet cousins circulated to the Royal Observatory or to Sharp in Yorkshire.[38] Jane Squire (bap.1686–1743) (whom we meet in Chapter 5) wrote a series of letters to a number of the Commissioners in the 1740s to accompany and explain her printed proposal, and then published the series of resulting correspondence, to justify her case to a public readership.[39]

As this plethora of works made the longitude problem increasingly publicly visible, John Harrison and the Commissioners of Longitude also took to print, becoming part of Flamsteed's 'swarme'. Harrison and his supporters left behind an unusual quantity of paper records: manuscripts, pamphlets, drawings, and broadsheets,[40] which link his writing to contemporary productions from handbills to novels and engravings. They produced single-sheet memorials, like Purshall and Billingsley,[41] and, like Squire, also published sections of Harrison's correspondence with the Commissioners.[42] They published a series of increasingly lengthy and vociferous pamphlets in which Harrison criticised the Commissioners and argued for his right to the 'great reward'. This forced the Commissioners into print likewise, to make publicly available the product of their many years' financial support of Harrison.[43] The archives both of the Board of Longitude and of the individual Commissioners show their concern to engage with public discussion and ensure their presence on the printed page.[44] To a large extent, in fact, the Board existed only in these paper exchanges, having no fixed location or

---

35 Flamsteed, *Correspondence*, Letter 1,360, p.700.

36 Hobbs, *A new discovery* (1714); Billingsley, *Longitude at sea*; Purshall, *Essay at the mechanism*. The non-conformist preachers are probably Isaac Hawkins (and his brother) who published Hawkins, *Essay for the discovery*.

37 Purshall's handbill, in particular, advertised a pamphlet published nine years earlier. Purshall, 'To the Lords Commissioners'; Billingsley, 'Since the publication'; Blundel, 'Advertisement'.

38 Flamsteed, *Correspondence*, Letter 1,372, p.719.

39 Squire, *Proposal*.

40 Quill noted this in *John Harrison*, p.209.

41 Harrison, *Case* (c.1767) and *Case* (c.1770).

42 In, for instance, Harrison, *Account*.

43 *Principles*.

44 This is clear, for instance, in the papers left by William Wildman, Viscount Barrington, surrounding the passage of the 1765 Act (Barrington Papers, Caird Library, BGN/1–14), and in papers of the clockmaker Alexander Cumming, employed by Parliament to adjudicate on Harrison's disclosure (Harrison Papers, CLC/L/CD/E/041/MS03973).

attendance, or even a secretary until 1763. Its contemporary existence was only in its paper products, those it received and those it produced.

The Commissioners' printed exchange with Harrison culminated in his book-length *A Description Concerning such Mechanism as will Afford a Nice, or True Mensuration of Time* (1775), which raises the complex question of who wrote Harrison's pamphlets. The voice is markedly different from the previous works printed under his name. It is verbose, angry, and confused, compared to its carefully structured and reasonably voiced predecessors, suggesting that Harrison's earlier works may well have been ghost-written.[45] Certainly, it is clear that both Harrison's brother James (1704–66) and son William (1728–1815) worked with him at different points in his career and represented him at the Board of Longitude.[46] One unpublished manuscript source, known as the 'Harrison Journal', is particularly important. This account of the Harrisons' interactions with the Board reproduces almost every letter, resolution, and conversation that took place between 1761 and 1766, and thus offers an invaluable insight into interactions not recorded in the Board minutes. It was presumably intended for publication, to put the Harrisons' case to the public. The 'Journal' exists in three manuscript copies, including one showing annotations in William Harrison's hand.[47] This hints at a collective process of authorship that further complicates Harrison's role in the longitude debates.

Such complex and non-transparent print productions were among the worries of another group that appears frequently in this book: the Scriblerians. Composed of the writers Jonathan Swift (1667–1745), Alexander Pope (1688–1744), John Gay (1685–1732), John Arbuthnot (1667–1735), Thomas Parnell (1679–1718), and the Lord Treasurer Robert Harley, Earl of Oxford (1661–1724), the group is named for their most famous production, *The Memoirs of Martin Scriblerus*, which created a fictional man of letters with whom to satirise the literary productions of their day.[48] By both

45 His fellow instrument maker and friend James Short is the usual candidate suggested by Quill in *The Man who Found Longitude*, pp.111–12. In his *Diary*, Jérôme Lalande stated that he was given a copy of the *Account* by Short, written by him with the lawyer Taylor White, Lalande, *Diary*, p.24.

46 James worked with John Harrison on early clocks in Barrow and originally accompanied him to London but returned to Yorkshire around 1739. See Quill, *The Man who Found Longitude*, pp.52–6. William Harrison is a constant advocate for his father in the Board minutes.

47 The copy with William Harrison's annotations is the main version used in this book, and is in a private collection. The Atkinson copy is in the State Library of Victoria, MS Box 110/9 H17809, the third in another private collection.

48 Kirby-Miller, *Memoirs*.

inventing absurd productions of Martin's pen and attributing to him existing contemporary works that they thought absurd, the Scriblerians highlighted their concern and distaste for the paper flood emanating from Grub Street. Proposals for finding longitude, as well as discussions and jokes about solutions, feature prominently in Scriblerian texts (as Chapter 3 particularly discusses). They worried both that the quantity of printed output diluted the morally improving works of which society stood in need, and, likewise, that readers would be unable to distinguish good from bad.

Hogarth saw the hack engravers who pirated his works as of the same ilk, showing similar concerns to the Scriblerians over the direction of a society awash with so much print. Both espoused a 'middle way', a need for moderation that likewise became central to the Scriblerians' main literary heir, Laurence Sterne's (1713–68) novel *Tristram Shandy*, which satirised the entire process of constructing literature (and which features in Chapter 6). Yet all of these writers also saw the danger that to criticise and engage with this paper world they too had to join the 'swarme' of authors and publications. Their constant claim, both serious and satirical, that all authors but themselves were seeking to impose on the public, was also the claim of every longitude pamphleteer (as Chapters 3 and 4 particularly show).[49]

Looking for Longitude

The following chapters pursue three routes through this mass of print culture surrounding the longitude problem. Each allows us to understand better how John Harrison came to such grief in his interactions with the Board of Longitude, and each brings us briefly back to William Hogarth. Part I considers longitude as a visual problem, focusing on the maps, diagrams, and words that were visual vehicles of knowledge in the longitude debate. Chapter 1 considers the process of defining longitude in dictionaries, and of agreeing precisely what the 1714 Longitude Act defined as necessary to win the reward. The problem of pinning down a line of longitude on a map is discussed in the context of different map projections, in relation to a line of magnetic variation, and within discussions by the Commissioners on the problem of establishing accurate coastal longitudes. Chapter 2 turns to the difficulties of illustrating instruments on paper, focusing on two key inventions in the longitude debate: the marine quadrant and the timekeeper. Processes of

49 This satirical position is discussed by Stallybrass and White in *Politics and Poetics*, p.104, and also explored by Lynall specifically in relation to longitude in 'Scriblerian Projections'.

patenting and copyright tie this into Hogarth's concerns surrounding his own engravings, and Harrison's difficulties in inscribing his invention on paper in a manner that the Board of Longitude could understand, as well as broader discussions of authorship and invention.

Part II looks at longitude as a mental problem, considering the language used to claim knowledge of solutions, and how the credibility of the author was assessed, questioned, and satirised in a wider context. In Chapter 3, projecting is compared to proposing a scheme, and projectors of longitude solutions offer the means to discuss wider concerns with projecting in society, from financial and political, to religious and literary. We see how the credibility of longitude ideas balanced on a fine line between madness and genius in Chapter 4, where the mathematical ideas that engrossed many longitude projectors could make them both mad and visionary, unable to communicate to the wider public. Yet a solution to the problem would, nonetheless, require a new association of ideas. The Scriblerian group of writers particularly feature in these discussions, which crucially affected Harrison's career, as well as Hogarth's.

Part III focuses on longitude as a social problem, turning from language to action, and the different visual and physical spaces in which longitude could be experienced. In Chapter 5 we see how discussions of, and solutions to, the longitude problem formed part of the polite sociable round of visits, images, and debates in which women provided a refining influence on men. We follow Hogarth as he visits Harrison's instruments as part of the round of London attractions. Yet longitude also featured in a genre of smutty contemporary poetry, which equated female intellectual empowerment with sexual licence, or latitude. Instruments are discussed as a means of both measuring and controlling such licence in Chapter 6, and this social tension becomes visible in Harrison's struggles to negotiate in an appropriate manner with the Board of Longitude, revealing his instruments and his knowledge in the right way. The problem of longitude emerges from these discussions as one of perception, of knowledge-making and of social expression. Looking for your longitude in eighteenth-century London required you to navigate society, as much as the open sea.

I

The Visual Problem

'the lines, which are so very fine'

> [besidez] the buoy's latine, him understandz something ov 'lozofy ... ant zayeth, that if him had but Littletonz dickzionary and Esops fables with cutts, him believeth that him could find out the longitude.[1]

Elements of these two chapters appeared in *Deutsche Gesellschaft für Chronometrie* (2013), pp.27–36.

1 *Grub Street Journal* 199 (18 October 1733), p.2. 'Cutts' here means woodcuts of accompanying images.

1

Naming and Mapping

HOGARTH'S IMAGE is unusual in the visual history of longitude as it takes a satirical approach to the problem and is more concerned with the figure of the lunatic than with his solutions. For most protagonists in the longitude debate it was the solution that was crucial, which required complex discussions around the meaning, veracity, and security of words and images on paper. Dictionaries and encyclopaedias changed markedly over our period, developing forms of knowledge organisation, in word and image, that changed the very definition of longitude as a word and a problem. It became rooted in particular processes of giving lines authority on paper, whether those be longitude lines on a map or the engraved image of a newly invented instrument. What was at stake was the question of how to create an image and a description of solutions that would win allies, not at sea, but in London, among those with the power to back proposals. Those solutions also needed to be tied securely to their inventor if he (or she) were to be sure of gaining the 'great reward'. Over the course of the century, discussions around copyright in printed texts, printed images, and patented instruments developed alongside the longitude problem, and involved Hogarth as well as Harrison.

'two terms of Art called Latitude and Longitude'[2]

The word 'longitude' might seem a simple way into understanding what this problem meant in the eighteenth century, but dictionaries and encyclopaedias were themselves changed over the period, as ideas about classification of knowledge developed. Both were part of new types of reading, looking, and knowledge seeking, as dictionaries, encyclopaedias, and atlases developed side by side. Their categories and arrangements provide us with a means to understand the contemporary view of knowledge, and therefore the structure of the world more broadly.[3] This also formed part of a general trend towards interest in, and application of, natural philosophical principles of standardisation.[4] 'Technical' language became open to everyone through increasing inclusion in such works, for which longitude is a prime example.[5] Likewise, dictionaries and encyclopaedias saw a growth in the use of illustrative figures over the eighteenth century. These allowed different modes of communication, and the condensing of complex ideas into a single image, but also posed problems of arrangement, labelling, and printing.

The huge expansion of interest in such publications precipitated a flood of dictionaries and encyclopaedias in eighteenth-century Britain.[6] High demand and turnover meant that works often lifted sections wholesale from previous publications and remained in print for long periods. One of the most famous, Nathaniel Bailey's (d.1742) *An Universal Etymological English Dictionary*, for instance, was in print constantly from 1721 to 1802. Yet definitions would be recycled only so long as they remained current, such that older dictionaries show a more rural way of life in their choice of words, whereas newer ones brought an urban focus.[7] It is significant then, that the definition of the word 'longitude' changed dramatically over our period, as understanding of what the word itself meant was changed through the process of trying to solve the problem. How this definition developed helps to show us what most concerned contemporaries in the process of giving it authority on paper. The Longitude Act, and proposed solutions to it, slowly became the definition of longitude, limiting the word to location at sea and the problem of finding that. Subsequent dictionaries saw the introduction of nuanced discussions of the different methods proposed so far. By the 1770s, John Harrison and

2 Newton, *Art of Navigation*, p.4.

3 Burke, *Social History*, p.94.

4 McIntosh, 'Eighteenth-Century English Dictionaries', p.7.

5 McIntosh, 'Eighteenth-Century English Dictionaries', pp.12–18.

6 McDermott, *Ashgate Critical Essays.*

7 McIntosh, 'Eighteenth-Century English Dictionaries', pp.3–6.

his work on timekeepers had become the definition. If classifications of knowledge can be read as expressions of the contemporary world view, then the changing definitions of longitude show us perfectly the changing status of longitude as a problem with wider implications.

Longitude's most basic meaning, length, generally constituted its first definition in early eighteenth-century dictionaries, followed by definition in terms of length of place, whether on land or sea. In 1702, for instance, 'J. K.' published his *New English Dictionary: Or, Compleat Collection of the Most Proper and Significant Words*. J. K. is assumed to be the John Kersey (fl.1702–20) who updated the sixth edition of Phillips' *New World of Words* in 1706 and was the son of the mathematician John Kersey (1616–77).[8] His dictionary stated that 'Longitude (in Geography) shews the distance of one place from another Eastward; to be reckon'd from some certain and fixed meridian, or line drawn from the North to the South Pole'.[9] The first really technical dictionary to appear was John Harris' (1666–1719) *Lexicon Technicum* in 1704. Harris was a mathematician and Fellow of the Royal Society who was able to draw on a wide range of his colleagues' works, as well as the advice of merchants and navigators.[10] In his preface, Harris emphasised how he wanted 'to make a Dictionary not only of bare *Words* but *Things*'.[11] He introduced the first specifically navigational definition of longitude, describing it as 'also the Distance of a Ship or Place, East or West from another counted in proper Degrees; but if in Leagues or Miles, or Degrees of the Meridian, and not in those proper to the Parallel of Latitude, it's commonly called Departure'.[12] He highlighted how, at this point, longitude and departure were largely used interchangeably by mariners, while he also included definitions of longitude 'in the heavens', 'of the sun or star from the equinoctial point', and 'in dialling', and an example of how to find the longitude and latitude of any star by the globe. Thus, longitude was most referenced as a term in astronomy. Navigation, by contrast, the preface explained, was 'largely treated of, and the whole *Art* taught under the Words *Plain* and *Mercator's Sailing, and Traverses*'.[13] This associated it directly with questions of map projection to which we will return. Harris'

---

8 On Kersey see Heddesheimer, 'The Authorship of *A New English Dictionary*' in McDermott, *Ashgate Critical Essays*, p.59. On his parentage see Wallis, 'Kersey, John, the elder' in *ODNB*.

9 Kersey, *New English Dictionary*, Q3r.

10 Bradshaw, 'John Harris's *Lexicon Technicum*' in McDermott, *Ashgate Critical Essays*, pp.231–43.

11 Harris, *Lexicon Technicum* (1704), A2r.

12 Harris, *Lexicon Technicum* (1704), 5C4v.

13 Harris, *Lexicon Technicum* (1704), A3v.

second edition of 1710 maintained this definition, and Bailey's *Universal Etymological English Dictionary* used a similar but shortened version in 1721, describing longitude as 'the Distance of a Ship or Place East or West from another, counted in proper Degrees'.[14]

By the third edition of Harris' *Lexicon Technicum*, published posthumously in 1723, however, the definition of longitude had changed dramatically. Longitude of a place had become the only version to be mentioned and the *Lexicon* now introduced a short discussion of the different methods of finding it at sea, emphasising the importance of this by explaining, 'the Difference of Time being turned into Degrees will truly give the *Longitude*, or *Vice Verso.* Hence several Ways have been thought of to find the Longitude at Sea; the great *Desideratum* of the Art of Navigation'.[15] It went on to suggest the use of lunar distance or Jupiter's satellites, highlighting the need for accurate tables but also the problem of using a large-enough telescope to observe at sea. Harris considered the importance, nonetheless, of these methods on land, stating that 'if Ships were sent with good Instruments and Men that know how to use them, to do this at all the Capes and Head-lands of the World, it would be a Thing of the greatest Use … by settling the Longitude of all those Places'.[16] He then went on to discuss the use of a clock to find longitude, and the problem of making one work reliably at sea. He ended by pointing the reader to the extensive entry on Mercator's sailing, which from the 1704 edition had featured a long entry on map projections and their practical use in navigation. The 1723 edition made the link explicit for the first time in the dictionary entry itself, rather than the preface.

Ephraim Chambers (c.1680–1740) next altered these definitions in 1728 in the *Cyclopædia.* He too became a Fellow of the Royal Society, after working as apprentice to the mapmaker John Senex (1678–1740).[17] Again longitude of a place was the only definition featured, but Chambers jumped almost immediately to its significance at sea, and introduced the influence of the 1714 Act and reward.[18] After an explanation along the basic lines continued from 1702, he continued, 'to discover an exact Method of finding the *Longitude*, especially at Sea, is a Problem, that has extremely perplex'd the Mathematicians … and for the Solution whereof, great Rewards have been publickly offer'd'. He went on to comment on the plethora of proposals, saying, 'Various are the attempts that Authors have made for this

14 Bailey, *Universal Etymological English dictionary*, Sss2r.

15 Harris, *Lexicon Technicum* (1723), ii, Kkk2r.

16 Harris, *Lexicon Technicum* (1723), ii, Kkk2r.

17 Bradshaw, 'Ephraim Chambers's *Cyclopaedia*' in McDermott, *Ashgate Critical Essays*, pp.247–61.

18 Chambers, *Cyclopædia*, ii, pp.470–2.

Purpose … but still without Success; all their Schemes being found either false, precarious, [or] impracticable'. Chambers also consciously organised his entries, switching the order of Harris' discussion to start with solutions proposing clockwork and explaining the difficulty of using a pendulum at sea.[19] He then listed all of the possible celestial methods with explanations of how to use them, but noted that the limitation here was the infrequency with which some phenomena appeared. Chambers ended by discussing Whiston and Ditton's proposal of firing shells at sea, noting that this was theoretically possible but not practical, then briefly referenced Whiston's scheme for using magnetic variation, pointing readers to the entry for 'Dipping Needle'. Elsewhere, under the entry on 'Pointing' he touched on the concern for marking longitude accurately on maps, explaining that 'all the difficulty in pointing a chart arises from our ignorance of the Longitude'.[20] Between them, Harris and Chambers firmly connected the definition of 'longitude' not only to its status as a problem at sea, but to questions of mapping, and to the observational and instrumental practices that might solve it on paper.

Samuel Johnson (1709–84), famous man of letters, nicely distilled these concerns in his contribution of 1755–56, *A Dictionary of the English Language: In which the Words are Deduced from their Originals, and Illustrated in their Different Significations.* The entry on longitude is a good example of how he did precisely what his title promised. He gave five definitions that built incrementally on each other, using well-known literary sources.[21] He started with definitions of longitude as length, and as the distance from a meridian, then added 'The distance of any part of the earth to the east or west of any place', using a quotation from John Arbuthnot's *Essay on the Usefulness of Mathematical Learning*. He then used Arbuthnot and the poet John Donne to define longitude in terms of two possible methods, firstly 'Of longitudes, what other way have we, / But to mark when and where the dark eclipses be? Donne', (*Valediction of the Book*) and secondly, 'His was the method of discovering the longitude by bomb vessels, Arbuthnot' (*The Memoirs of Martin Scriblerus*). Thus, Johnson not only continued the emphasis on methods for establishing longitude at sea, but he also referenced precisely the much-discussed methods of eclipses and projectiles that Hogarth's lunatic is drawing on the wall of Bedlam. We will return in Chapter 3 to the further significance of Johnson's choice to use Arbuthnot's comment on Whiston's 'bomb vessels'. Meanwhile, the poet and periodical writer, Richard Rolt

19 Bradshaw, 'Ephraim Chambers's *Cyclopaedia*', p.250.
20 Chambers, *Cyclopædia*, ii, p.844.
21 Johnson, *Dictionary*, ii, 4D4v.

(1724–70), was paying a little more attention to occurrences at the Board of Longitude and in Parliament, which allowed him to update the story of longitude in his *New Dictionary of Trade and Commerce*, which appeared the same year as Johnson's second volume.[22] Rolt largely followed Harris' latest definition, but added the details of the 1740 Act, which had expanded the terms of 1714 to allow for payments to survey the British and Irish Coasts. He was, therefore, able to reference the money given in 1741 to John Harrison for his timekeepers, and to William Whiston for surveying the coasts using magnetic variation.[23] Thus, both Harrison and the process of regular funding awards by the Commissioners became part of the dictionary definition of longitude.

By 1763, in the *New and Complete Dictionary of Arts and Sciences* from the 'Society of Gentlemen', Harrison was playing a prominent role in the definition. After the basic explanation of longitude as distance, a list of the relevant acts, and brief details of the potential celestial methods, this dictionary went on to explain how timepieces might succeed. It stated that, 'the most ingenious and successful machines for this purpose have been invented by Mr. John Harrison, who, at different times, contrived three different time-pieces for determining the longitude at sea'.[24] It went on to give a history of his three timepieces, which mistakenly elides Harrison's third and fourth clocks into one. It ended with the details of the Jamaica trial and the clock's successful performance, concluding that this amounted 'to a demonstration, that Mr. Harrison has performed all that is required by the statute of the 12th of queen Anne'. Here we see the first direct relationship being drawn between the performance of Harrison's timekeepers and the exact terms of the 1714 Act: a relationship on which, we shall see, Harrison himself also placed much store. Elsewhere, the dictionary built on Chambers' and Rolt's brief discussions of magnetic variation, explaining Whiston's theory of magnetic dip. It explained how 'to find the longitude or latitude by the dipping-needle. If the lines of equal dip below the horizon be drawn on maps or sea-charts from good observations, it will be easy … to find the longitude either at sea or land', but also discussed how, 'About the year 1744, Mr. Mountaine and Mr. Dodson found, on a careful examination, that the variation lines constructed by Dr. Halley were become entirely useless'.[25] This highlighted the problem of pinning longitude lines down, which obsessed a

22 Betty Rizzo, 'Richard Rolt' in *ODNB*.

23 Rolt, *New Dictionary*, 612–612v.

24 Society of Gentlemen, *New and Complete Dictionary*, iii, pp.1,968–70, quotation p.1,970.

25 Society of Gentlemen, *New and Complete Dictionary*, iii, pp.2,232–4, quotations p.2,233.

number of our protagonists, and which features later in this chapter. In 1771, the first edition of the enduring *Encyclopaedia Britannica* appeared, with the clearest statement of how very specific the word 'longitude' had become since 1700. The *Encyclopaedia* largely copied the 1763 *New and Complete Dictionary*, 'Longitude of a place' was again the only entry, but it cut any explanation of distance in relationship to a meridian. Instead, the entry started immediately with the information that, 'In the philosophical transactions, no 1, we have an account of a successful experiment in finding the longitude at sea', and went on to discuss solely the development of clockwork, focusing swiftly on Harrison.[26] Whiston's name no longer featured in the brief mention of magnetic variation at the end. Thus, the definition of longitude moved out of our period as entirely a question of place at sea, of proposals to find that, and of Harrison's timekeepers.

'THE ACT OF QUEEN ANNE'

HARRISON TOO was concerned with definitions; we might say obsessed. During his nearly forty-year interaction with the Commissioners of Longitude he constantly relied on, and referred back to, the particular terms of the 1714 Act. Much of Harrison's later acrimonious discussion with the Commissioners was over defining how his watch could be deemed to have shown itself 'practicable and useful at sea'. He was well aware of how important the definitions of the terms of the Act were, and constantly tied his claims and complaints back to these. Wrangles over these definitions started in 1763, after William Harrison had returned from the first trial of the watch to Jamaica, and the Harrisons had disagreed with the Board over the outcome. Harrison was encouraged by friends to apply to Parliament for an Act to reward him with five thousand pounds for disclosing the workings of his watch, and to give him exclusive right to the reward over any other timekeepers for the next four years. To give him, effectively, copyright on the use of a clock to find longitude. In the petition that he submitted to the House of Commons for that Act, Harrison commenced with the observation 'That by Statute 12th of Anne a Reward of £20,000 was enacted for the discovery of the Longitude at Sea. That an Instrument which could keep true Time at Sea … has been allowed on all hands to be the best method of discovering the Longitude for the use of Navigation'.[27] He stated clearly the terms of the Act, and the belief that a watch could do this. At the same

26 *Encyclopaedia Britannica*, ii, pp.1,004–5.
27 *HJ*, p.37.

time, the Harrisons published *An Account of the Proceedings, in Order to the Discovery of the Longitude*, in which they asserted clearly how their watch had fulfilled the terms of the Act. They explained how the results of the trial amounted to

> A Demonstration, that Mr. *Harrison* had performed all that is required by the Statute of the 12th of Queen *Anne* ... and consequently that whatsoever further Experiments have been made, or may hereafter be made ... ought to be without Prejudice, to the Right he has acquired, by Virtue of his Compliance with [the] Act.[28]

This continued after the second trial to Barbados in 1763–64, as the Harrisons and the Commissioners again attempted to agree on what the Act required. In all of his discussions with the Board, Harrison showed the same dogged reliance on the wording, opening one petition in 1766 by stating, 'The Act of the 12th of Queen Anne (upon the faith of which I have worked 40 years) sets forth ...'.[29] Furthermore, in a number of publications, Harrison started to comment on the legality of the Commissioners' right to withhold the reward. In *The Case of Mr. John Harrison* in 1767, he asserted how, 'On Consultation ... with several eminent Men of the Law, ... [he] found them unanimously of Opinion that he was *entitled* to the greatest Reward under the Act of Queen *Anne*'.[30] Similarly, in *Remarks on a Pamphlet Lately Published by the Rev. Mr Maskelyne*, he complained how the Commissioners now had discretion over further trials: 'an Exposition of the Law, which I ever did and ever shall (until it is supported by legal Authority) totally reject and refuse Obedience to'.[31] The Act had prioritised practice over theory, emphasising how a scheme needed to be proved 'practicable and useful at sea', not merely a theoretical possibility. Yet what these arguments in 1765 showed was that paper was still important; definitions and solutions still needed to be agreed in writing. This is amply shown by the small archive of papers from the passage of the 1765 Act that were kept by William Wildman, Viscount Barrington (1717–93), then a Commissioner of Longitude in his role as Treasurer of the Navy.[32] In his speech to Parliament explaining why the Commissioners were seeking the new Act, he explained,

> If the Commissioners had been certain that other timekeepers could be made which would enable other Ships to find their Longitude with equal exactness

28 Harrison, *Account*, p.11.
29 *HJ*, p.107.
30 Harrison, *Case* (c.1767), p.2.
31 Harrison, *Remarks*, pp.20–1.
32 Discussed in Barrett, '"Explaining" Themselves', pp.145–62.

> they would have given a Certificate for the great reward. But no other Watch has been made and they know not the principles of this [one]. They think therefore they are not justified according to the Act to give the Certificate till farther discovery be made in these points.[33]

They needed clarification on paper in order to give the reward 'according to the Act'. This meant not only that Harrison needed to 'discover' the watch in such a way that it could be understood by other makers but, crucially, that copies needed to be made by these makers as well as by Harrison.[34] These 1765 disagreements were, essentially, an attempt to clarify the definitions of 1714 in largely the same way that the dictionaries did over the same period.

'the lively representation of the Position of Places'[35]

We have seen how dictionaries linked longitude to a range of entries dealing with maps and charts. John Harris joined longitude to his discussion of the Mercator projection in the 1723 *Lexicon Technicum* and emphasised from 1704 in the opening preface how 'as to the *Variation of the Compass*, I have given you a very full and exact Account of it from that Excellent Mathematician Capt. *Halley*'.[36] We now turn to consider this mapping concern. How could you go about plotting the longitude of a place, or a line of longitude, that 'distance of one place from another Eastward; to be reckon'd from some certain and fixed meridian'? A new English translation of Bernhard Varenius' (1622–50) seminal seventeenth-century work *Geographia Generalis* appeared in 1733.[37] In this work, Varenius used his definition of longitude to discuss the problem of pinning it down on a map, explaining how 'there is no Meridian determined in Nature to be the first, and therefore any one may be taken'.[38] He suggested that the words 'longitude' and 'latitude' in fact confused understanding of how such lines worked, saying, 'properly speaking, a Place or Point, hath neither Latitude nor Longitude: and these two different Sense of the Words ought to be well remembered … in my Judgement it were better not to use the Words Latitude and Longitude, but

33 Barrington Papers, Caird Library, MS BGN/8, 1v.
34 Bennett, 'Travels and Trials', pp.75–95.
35 Newton, *Art of Navigation*, p.61.
36 Harris, *Lexicon Technicum* (1704), A3v.
37 First published Amsterdam, 1650. Newton produced a re-worked Latin edition in 1672, which was updated by James Jurin in 1712. This is the version that Dugdale translated into English in 1733.
38 Varenius, *Compleat System*, ii, p.655.

only Distance from the Equator, and from the first Meridian: but being so long used they cannot be now abolished'.[39] He saw the danger of relying on latitude and longitude as easy units of measurement, which the words themselves encouraged.

The problem for contemporaries was two-fold. Firstly, mentally to transfer the grid of lines on a globe onto the world, anchoring these to reliable reference points; and secondly, to agree which points or lines should be taken as the departure point on a round globe. This involved not only agreeing a common first meridian (which would not be achieved until 1884) but also establishing the longitude of points on land with agreed accuracy. In his pamphlet published soon after the passage of the Act in 1714, *A Practical Method, To Discover the Longitude at Sea*, the Chester mathematician and educator, John Ward (c.1679–1758) recognised these problems by recommending to mariners that they select rooms where meridian lines could be drawn permanently on the ceiling using reflections of the sun's rays.[40] In essence, he sought to create permanent longitude lines, anchoring them in one place. In another pamphlet from 1714, among a discussion of the proposals that he thought impossible, the inventor and financial adventurer Case Billingsley commented that 'we may with as much reason hope e're long to see the Heavens drawn with curious Black Meridians through every tenth degree of the Equator'.[41] He made the problem of establishing longitude lines beautifully visible. Ward and Billingsley's contemporaries went on to consider two main systems of lines that might stabilise these concerns, both of which we have already seen referenced in the dictionaries. One was the introduction of the Mercator projection for sea charts, as a means of ensuring that the grid of latitude and longitude was drawn on maps in such a way that it could easily be used to navigate. The other was the mapping of a second set of lines, of magnetic variation, onto these maps, which the navigator could follow simply through use of a sensitive compass. Longitude thus led fundamental shifts in the period around how images helped to establish objectivity and reliable forms of reference for scientific ideas.[42]

---

39 Varenius, *Compleat System*, ii, pp.256–7.

40 Ward, *Practical Method*, pp.21–4.

41 Billingsley, *Longitude at sea*, p.11. See Rogers, 'Gulliver and the Engineers' in *Eighteenth Century Encounters*, p.16, and Cummings and Stewart, 'The Case of the Eighteenth-Century Projector', pp.235–61.

42 See, discussions in Baigrie, *Picturing Knowledge*; Pauwels, *Visual Cultures*; Jones and Galison, *Picturing Science*; Daston and Galison, *Objectivity*; Edgerton, 'From Mental Matrix to Mappamundi' in Woodward, *Art and Cartography*, pp.10–50; Black, *Maps and History*; Turnbull, *Masons, Tricksters and Cartographers*; Latour, *Pandora's Hope* and 'Visualisation and Cognition'.

'of Mercator's chart'

The process of projecting longitude and latitude lines from a globe onto a flat surface required prioritising accuracy of either shape or size. As the longitude reward was being announced and discussed in London coffee houses in the early decades of the eighteenth century, mariners were slowly accepting the Mercator projection as the standard navigational chart. The Flemish cartographer Gerard Mercator (1512–94) had presented his cylindrical map projection in 1569, suggesting a new means of representing the spherical earth on a flat surface. Mercator's ideas, too complicated for ordinary navigation, were later made intelligible by the Englishman Edward Wright (1561–1615) in his 1599 publication *Certaine Errors of Navigation* and the accompanying world chart of 1600.[43] Wright and Mercator's combined work was ground-breaking in allowing navigational courses to be plotted as straight lines on the projection, maintaining their angle with lines of longitude. This made the projection much easier for setting a course at sea (although harder for calculating distance). Yet Wright's comparative obscurity as the author of the projection compared to Mercator was something that many English writers bemoaned. One such was Samuel Newton (fl.1695–1708), Master of the Mathematical School at Christ's Hospital, in his *Art of Navigation*, published the year after the Longitude Act passed. After stating that it should more properly be called Wright's chart, he outlined how this projection improved practical sailing:

> All other Maps, whose Construction depended upon the Optic projection of the Sphere, were (even the best of them) unfit for the Use of ordinary Seamen, for though the Latitude and Longitude of Places, might be truly laid down upon them, yet the bearing and distance of Places (being of greatest use to Seamen) could not possibly be found in some of them … thereupon [Mercator] applied himself to the discovery of such a Construction as might not only shew the Latitude and Longitude of places, but also their bearing and distance, in right Lines.[44]

He emphasised, in effect, how practical drawing on paper was one of the great successes of the new projection.

One of the publishers involved in producing many map projects in the eighteenth century was John Senex. He had moved into the map trade from being a stationer, but also engraved many maps himself, including works by Halley and Whiston, and became a Fellow of the Royal Society

43 Moreland and Bannister, *Antique Maps*, p.34.
44 Newton, *Art of Navigation*, p.75.

in 1728.[45] This was the same year in which his apprentice, Ephraim Chambers, produced the first volume of his *Cyclopædia*, in which he discussed the Mercator chart at some length.[46] Chambers ended with a brief mention of another projection: '[a] *Globular Chart*, is a Projection so call'd, from the Conformity it bears to the *Globe* it self; lately propos'd to the World by Messrs. *Senex*, *Wilson*, and *Harris*'. He justified this new projection and dismissed one protagonist in particular, saying,

> This Projection is yet in its Infancy; and the Secret thereof in a few Hands: We have but little to say therefore as to its Merits or Failings ... The Cavillings of Mr. *Haselden*, who has thought fit to censure it e'er he knows what it is, will weigh but little.[47]

In advertising the latest projection of his former master, Chambers thus glossed over an extended and detailed argument that took place between Henry Wilson (1673–1741) and Thomas Haselden (d.1740) in 1722, over how best to draw longitude lines on a map. They discussed specifically the merits of different projections for plotting longitude lines and, crucially, the actual process of drawing these. The first surviving entry into the fray came from Haselden in response, it seems, to a sheet published by Wilson in 1718.[48] Wilson's sheet is no longer extant, but was likely similar to one in the resulting *Atlas Maritimus & Commercialis*, which would be published in 1728 (Figure 4).[49] Wilson announced his new projection, which he called 'globular' based on its exact representation of the globe, and which he claimed was more 'natural' and, therefore, more accurate than the Mercator projection and easier for use in navigation (Figure 5).[50] The announcement sought subscriptions for the *Atlas* based on the projection, which Wilson, Senex, and Harris also patented in 1721.[51] Haselden responded in defence of the Mercator projection and questioned the accuracy and purpose of the globular chart. He particularly cavilled at the support that Wilson and Senex claimed the globular projection had from Edmond Halley. A long pamphlet from Wilson—and another in reply from Haselden—then argued the merits of the different projections in ever increasing detail, and the questions of how each could practically be drawn and used.

45 Tyacke, *London Map-sellers*, pp.xii, 142; Laurence Worms, 'John Senex' in the *ODNB*.

46 Chambers, *Cyclopædia*, i, p.202.

47 Chambers, *Cyclopædia*, i, p.202.

48 Haselden, *Description and use*.

49 *Atlas maritimus & commercialis*.

50 Wilson's argument can be gleaned from his 1722 pamphlet *The description and use*.

51 'Making and printing globular charts or sea charts for use in navigation' in Woodcroft, *Alphabetical Index*, 432.

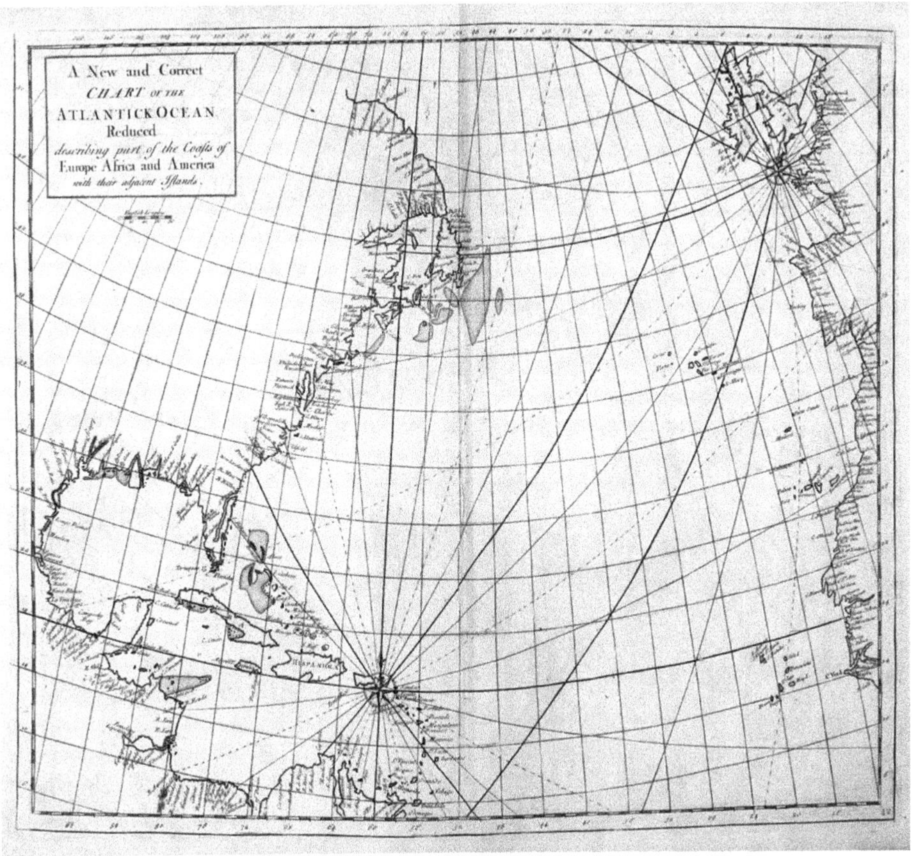

Figure 4: 'A New and Correct Chart of the Atlantick Ocean reduced describing part of the Coasts of Africa and America with their adjacent islands' from *Atlas maritimus & commercialis* (London, 1728); engraving, 510 x 590mm, Royal Museums Greenwich, London.

The key arguments involved two aspects of the different projections. The first concerned whether longitude lines could be accurately drawn. As Haselden emphasised, 'the main Question that relates to the Mercator's, or indeed to any other Chart, is whether it will show by a ready Method, the true Bearings and Distances of Places when laid down, according to their true Latitudes and Longitudes',[52] and thus far Wilson agreed, 'the finding the Longitude at Sea, remains so imperfect; and … we never, till now, had any Sea-Charts publish'd, that appear'd like the Surface of the Globe'.[53]

52 Haselden, *Description and use*, p.iii.
53 Wilson, *Description and use*, pp.5–6.

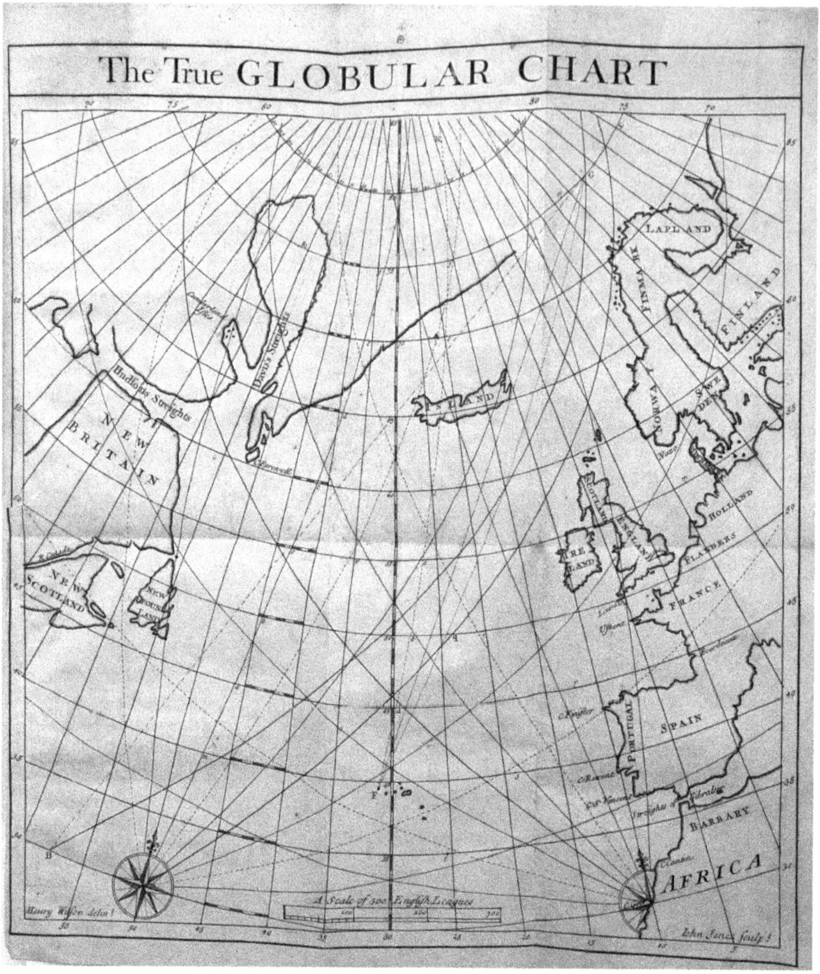

Figure 5: 'The True Globular Chart' from Henry Wilson, *The description and use of that most excellent invention, call'd the globular chart* (1722); engraving, 340 x 310mm, R.T. 52(8), © History of Science Museum, University of Oxford.

Wilson claimed that he had made the key improvement of projecting landmasses in their actual relationship to each other—both in size and distance. This meant that distances could be measured directly on his map with a scale of equal parts and enabled the 'great circles' of longitude to be properly drawn and measured. Haselden, in return, attacked the basic geometry behind Wilson's projection, saying, 'since most of the Triangles on this Chart consist of a Great Circle, a Small Circle, and a Spiral Line, the World will certainly expect a New Trigonometry, quite different both from

Plane and Spherical'.[54] Haselden contended further that the globular chart was false in its use of rhumb lines, and would therefore lead to shipwrecks,[55] and discussed at length how to construct a Mercator chart with particular emphasis on longitude lines. The key point was that the Mercator projection turned curved latitude and longitude lines into a parallel grid where, in the globular, they became circular.[56]

Haselden's point here was that Mercator's 'right lines' were much easier for navigators to draw, and it was secondly over the physical process of drawing their projections that he clashed with Wilson.[57] Each included both maps and diagrams in their pamphlets and questioned how the other had drawn both their own and the opposing projection. Haselden included sample diagrams in his first response (Figure 6), and his own version of the globular projection variously appeared in both of his pamphlets (Figure 7).[58] In his second response to Wilson, Haselden questioned 'you represent the *Globular Chart* as natural to the Eye; pray where do you fix your Point of Sight?'[59] and argued that no one could think that 'a Curve line may be drawn, and any Portion of it measur'd, and a Curvilineal Angle constructed with equal if not greater Exactness, and in less Time than a right [meaning straight] Line'.[60] His version of the globular chart emphasised the projection's problematic curved nature, using many fewer rhumb lines, and making the grid of latitude and longitude appear more substantial and dependable. Haselden even made the outline shape of the map echo the latitude–longitude grid, where Wilson's cuts a rectangle border across all of his lines. Haselden's chart, therefore, works to make the Mercator grid appear more natural, directly attacking Wilson's claims. Wilson, in return, repeatedly complained that Haselden had not followed his instructions and had drawn his projection incorrectly, saying, 'your drawing a false [projection] is not sufficient to prove ours to be so'.[61] Yet the point was not just their own projection in the pamphlets, but the practical ability to make and use these day to day. The argument ended with Haselden challenging Wilson, 'because this can't be so well and so effectually performed by SAYING, as it can by DOING, I am very willing

54 Haselden, *Description and use*, p.v.
55 Haselden, *Description and use*, pp.viii–ix.
56 Haselden, *Description and use*, pp.7–11.
57 Haselden, *Description and use*, p.16.
58 For instance, the map is bound into the National Maritime Museum's copy of *Description and use*, but not into the British Library copy. It appears in the British Library copy of *A reply*.
59 Haselden, *Reply*, p.24.
60 Haselden, *Reply*, p.vi.
61 Wilson, *Description and use*, p.37.

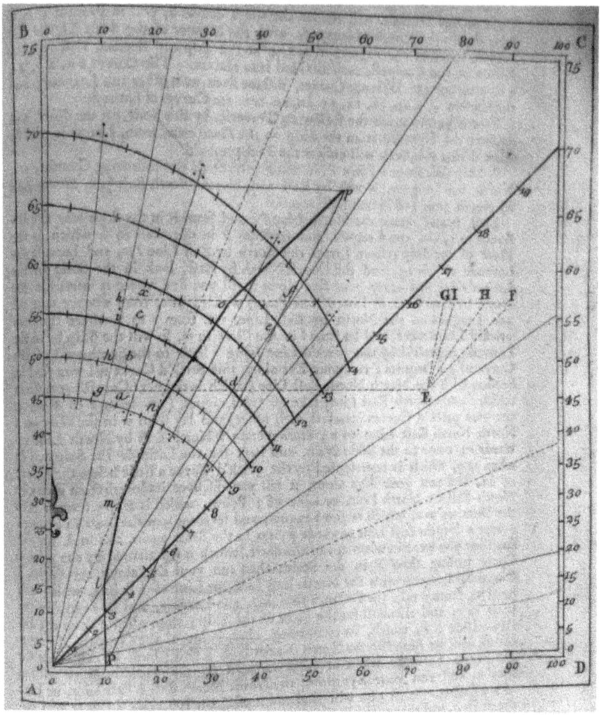

Figure 6: Diagram from Thomas Haselden, *The description and use of that most excellent invention commonly call'd mercator's chart* (London, 1722); engraving, 130 x 150mm, Royal Museums Greenwich, London.

to submit the decision of this controversy to an *Experiment*, to be made before the *Royal Society*'.[62] He proposed that each would bring clean paper, and proper tables and instruments and create a chart for the same longitude and latitude. Using these, they would then solve five navigational problems and draw one full traverse on their chart.[63] The two continued to argue over such an 'experiment' in the *Post Boy* and *Daily Post* and were critiqued in the *Weekly Journal*.[64] They did not compete at the Royal Society.

Neither Wilson nor Haselden seems to have engaged directly with the Commissioners of Longitude in this debate, or to have considered their projections as a candidate for the reward. The otherwise obscure William Blundel did however, addressing the Commissioners in a series of short pamphlets, in which he sought financial return for his projection, and criticised the monopoly that he saw Wilson and Haselden as having over public attention. In late 1723, Blundel published his pamphlets together in a

62 Haselden, *A reply*, p.29.

63 Haselden, *A reply*, pp.29–32.

64 *Post Boy* (1–3 November 1722), Issue 5,193 and (6–8 November 1722), Issue 5,195; and *Daily Post* (29 April 1723), Issue 1,118; *Weekly Journal* (15 June 1723), p.2,578 (as discussed in Chapter 4).

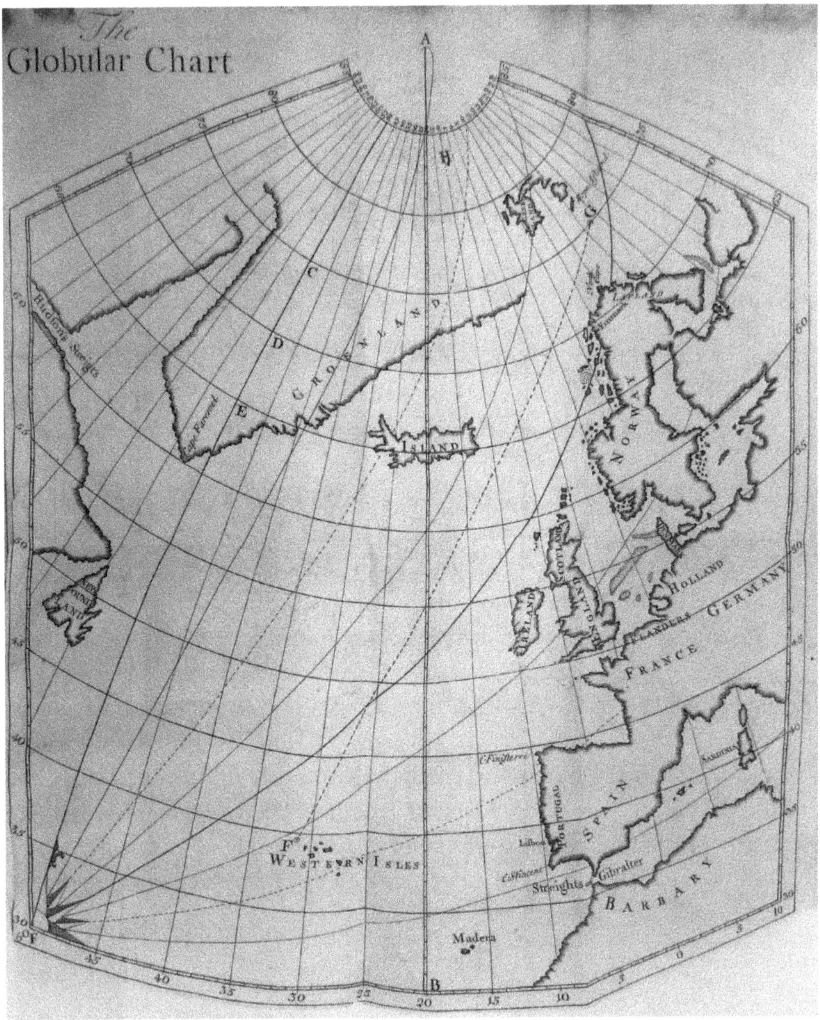

Figure 7: 'The Globular Chart' from Thomas Haselden, *The description and use of that most excellent invention commonly call'd mercator's chart* (London, 1722); engraving, 299 x 353mm, Royal Museums Greenwich, London.

set including contributions written during the previous two years.[65] In one from April 1723, he introduced a series of printed papers and instruments to

65 This is a bibliographically complex work. Blundel appears to have combined a number of publications into one. It has no title page so is catalogued in the ESTC based on the opening lines. pp.1–5 are dated 19 December 1721; pp.6–14, 29 September 1722; pp.15–26, 2 April 1723; pp.27–30 are undated.

the Commissioners that would 'put to silence those sharp disputes between Mr. *Willson* and Mr. *Hazeldine*'.[66] Later, Blundel argued for the accuracy of his own projection by referring to Wilson's criticisms of Mercator's chart. He advised the Commissioners to

> See Mr Willsons Book, of the Description and use of the Globular Chart … what a number of Inconsistencies there is there mentioned to be in Mercators Chart; but what hath Mr Willson, or Mr Hazeldine, or any other done to give apprehensions of the Inconsistincies not only of Mercators Chart, or any [other] Chart, whereby to refute or discover their Inconsistency; this ought to be done by demonstrations and proofs.[67]

Like Haselden, he argued for the need to demonstrate the projection's efficacy by drawing, but interestingly went on from this to comment that Wilson's projection did not approach anywhere near to the accuracy required by the 1714 Act.[68] He criticised Wilson, Haselden, and the Commissioners for attending to such disputes rather than to his publications: 'Were the Zeal of Mr Willson … so ardently affected to the publick Good, and Benefit, … [he] would leave off those hot disputes which proceed from Erroneous Suggestions on both sides' and would answer him, Blundel, instead. Blundel complained that his publications had appeared either side of Wilson's and Haselden's but had received no attention, even though he had tried to correct their errors. The Commissioners need simply read and apply his discoveries.[69] Blundel included no figures or diagrams in any of his publications, but both Wilson and Haselden understood the importance of visualising their projections in pamphlets as a means of demonstrating their accuracy. We will see how these concerns affected projections of invented instruments as well as maps.

'THE VARIATION LINES'

THE MERCATOR/WRIGHT PROJECTION was still being debated in the 1760s as Harrison's fourth timekeeper was sent on trial by the Board of Longitude. In 1762, John Rowe (fl.1751–62), accountant and author of *An Introduction to the Doctrine of Fluxions*, revised and published a posthumous work by the Reverend William West of Exeter (d.1760), entitled *Mathematics.*

66 Blundel, *To all and Singular*, pp.15–16.
67 Blundel, *To all and Singular*, p.28.
68 Blundel, *To all and Singular*, p.28
69 Blundel, *To all and Singular*, pp.29–30.

With Eleven Copper-plates. This again criticised Mercator's projection and suggested a different method of projecting a sphere onto a cylinder of paper.[70] This was, in turn, critiqued in the pages of the *Philosophical Transactions* by the mathematician Samuel Dunn (1723–94), who warned that 'Mr. West and his editor have ... fallen into this error; they have taken the words but not the sense of Edward Wright'.[71] He argued, effectively, that they should have paid more attention to the lines and less to the explanation of Mercator/Wright's projection. Another criticism came from the mathematician, teacher of navigation, and Fellow of the Royal Society, William Mountaine (c.1700–79). In his letter to the *Philosophical Transactions*, Mountaine placed West in the context of the earlier debates between Wilson and Haselden, outlining how, 'about the year 1720, a curvilinear sea chart made its appearance, said to be done by Henry Wilson, the publishers whereof represented Wright's chart as puzzling, difficult, and false. But these groundless assertions were rationally answered by Mr. Thomas Haselden ... in a letter and pamphlet addressed to Dr. Halley'.[72] Mountaine made a direct comparison of West and Wright on how to carry out the proposed projections, including a practical discussion of how to draw each, and concluded that 'by comparing these two methods of construction together I think it is not very difficult to discover that Mr. West's derives from Wright's'.[73] In effect, if Wright's were wrong then West's must be too, and vice versa.

In fact, in the period between Wilson's and West's critiques, Mountaine had been involved in creating a very different system of lines from which to find longitude. The problem of measuring longitude at sea articulated a fight about lines that stretched further than the simple grid. These 'Halleyan Lines' were designed to be mapped onto a Mercator projection but were an alternative to the grid of latitude and longitude. Edmond Halley, Flamsteed's successor as Astronomer Royal, and the patron of Wilson's projection, had published two charts showing lines of magnetic variation in 1701 and 1702, following his voyages on the *Paramore* (1698–1700). These built on the hypothesis that the variation between true and magnetic north in different parts of the globe could be mapped into lines, such that this variation could then be used to navigate. Lines of magnetic variation would, the idea was, create a measurable equivalent to a line of longitude. First, Halley published a chart of the Atlantic Ocean that he had navigated (Figure 8). This was

70 West, *Mathematics*.

71 Dunn, 'Remarks on the Censure of Mercator's Chart', read November 1762 in *Philosophical Transactions* (1763), pp.67–8.

72 Mountaine, 'A Defence of Mercator's Chart', read March 1763, *Philosophical Transactions*, p.72.

73 Mountaine, 'A Defence', pp.74–7.

Figure 8: Edmond Halley, *A New and Correct Chart Shewing the Variation of the Compass in the Western and Southern Oceans as Observed in the year 1700* (1701); engraving and letterpress, 580 x 490mm, Royal Geographical Society.

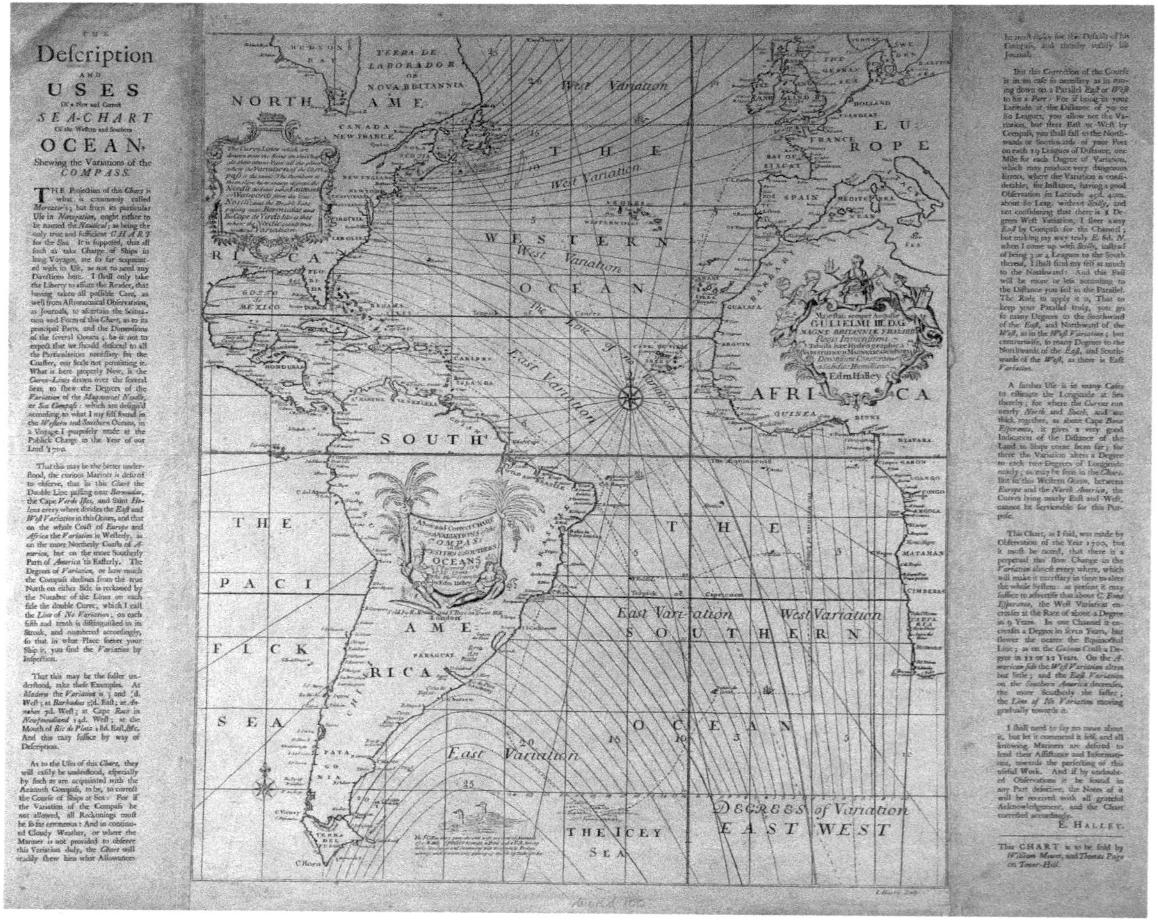

produced as a single large sheet, before an accompanying description became necessary, and was printed on two sheets to be pasted on each side.[74] As Halley there explained, 'what is here properly New, is the *Curve-Lines* drawn over the several Seas, to shew the Degrees of the *Variation* of the *Magnetical Needle*'.[75] His innovation was in creating a convincing pattern of global lines from a fairly sparse, and effectively disguised, set of observations.[76] We note how busy the seas have become, filled with curving variation lines of different thicknesses, as well as with dotted lines plotting Halley's two voyages and a central wind rose emitting rhumb lines. The edges of landmasses appear to be more the edges of the oceans, their interiors providing blank white space on the page that Halley has filled with explanatory inscriptions.[77] Likewise, the most southern part of the Southern Ocean is inhabited by sea monsters and islands of ice, placing the map in an older tradition of courageous exploration.

Halley followed this chart with a second one, of the whole world, which would become the influential version for navigators (Figure 9). This is somewhat simplified, having lost both the tracks of Halley's voyages and the rhumb lines, as well as the visually anecdotal sea monsters and 'Icey Sea'. Likewise, the cartouches and inscriptions have been altered and moved to different landmasses. Significantly, in this chart, Halley extended the variation lines to areas that he had not explored but for which he had reliable observations, such as the Indian Ocean.[78] The 'Great South Sea', however, remained a simple grid of latitude and longitude, except for a box of text in which Halley explained that 'I durst not presume to describe the like Curves in the South Sea wanting accounts thereof'. In both charts, he made clear the importance to how they should be read of the manner in which the lines were marked out on the map, as much where lines were not drawn as where they were. Halley's magnetic variation hypothesis continued to vie both with other longitude solutions and with other cartographic discussions throughout our period. In 1741, in the historical preface to his pamphlet *The Longitude Discovered by the Eclipses, Occultations and Conjunctions of Jupiter's Planets*, William Whiston commented that the hypothesis had potential but that Halley's map was badly in need of updating: 'by reason of this Variation of the Variation, it is very fit that a new Map … be made at least every ten Years; and placed in every Ship that

74 On Halley's two charts see Thrower, *Three Voyages*, pp.56–60.

75 Reproduced in Thrower, *Three Voyages*, p.365.

76 Thrower, *Three Voyages*, Figure 5 (facing p.48) plots Halley's observation points on a map making clear how much he extrapolated from his data.

77 Christian Jacob discusses such negotiation of space use on a map in *Sovereign Map*.

78 Jacob, *Sovereign Map*, p.60.

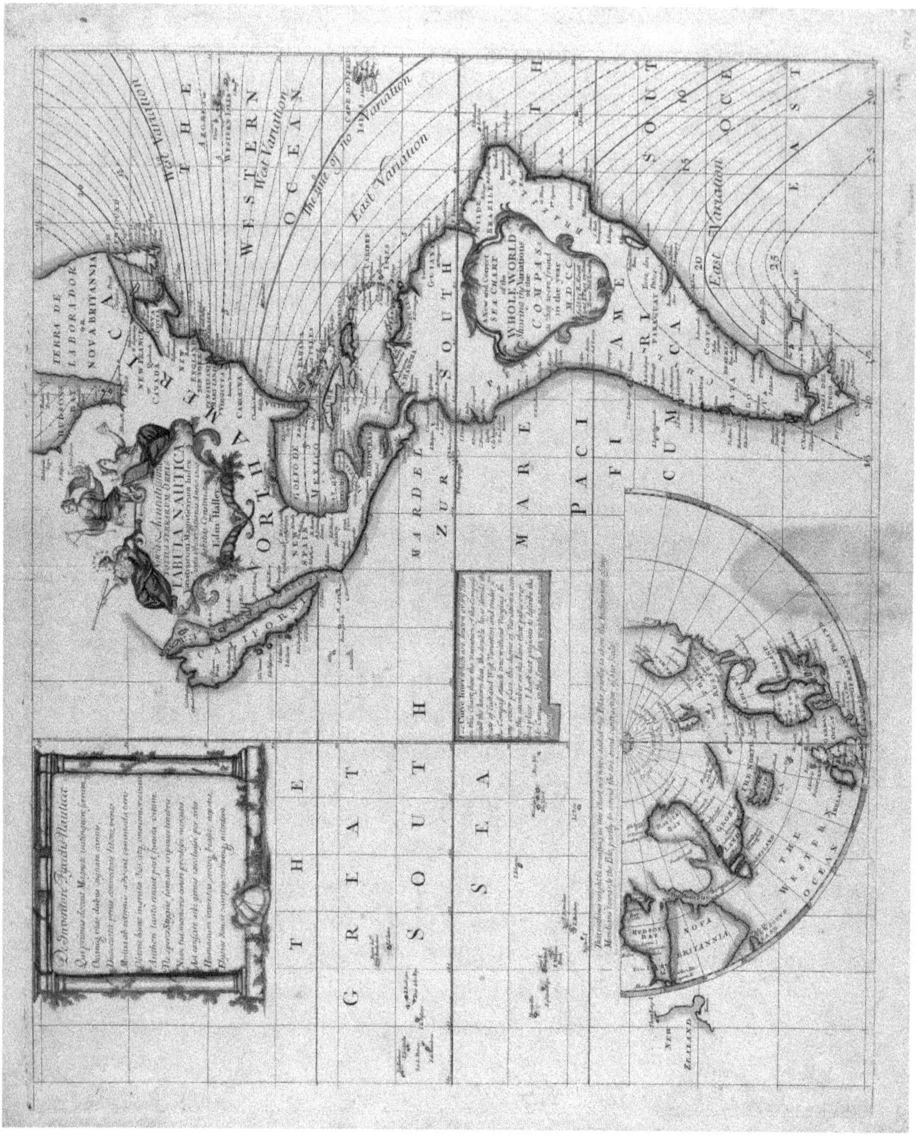

Figure 9: Edmond Halley, *A New and Correct Sea Chart of the Whole World Shewing the Variations of the Compass as they were found in the year 1700* (1702); hand-coloured engraving, 570 x 1,380mm (in two sheets), Royal Museums Greenwich, London.

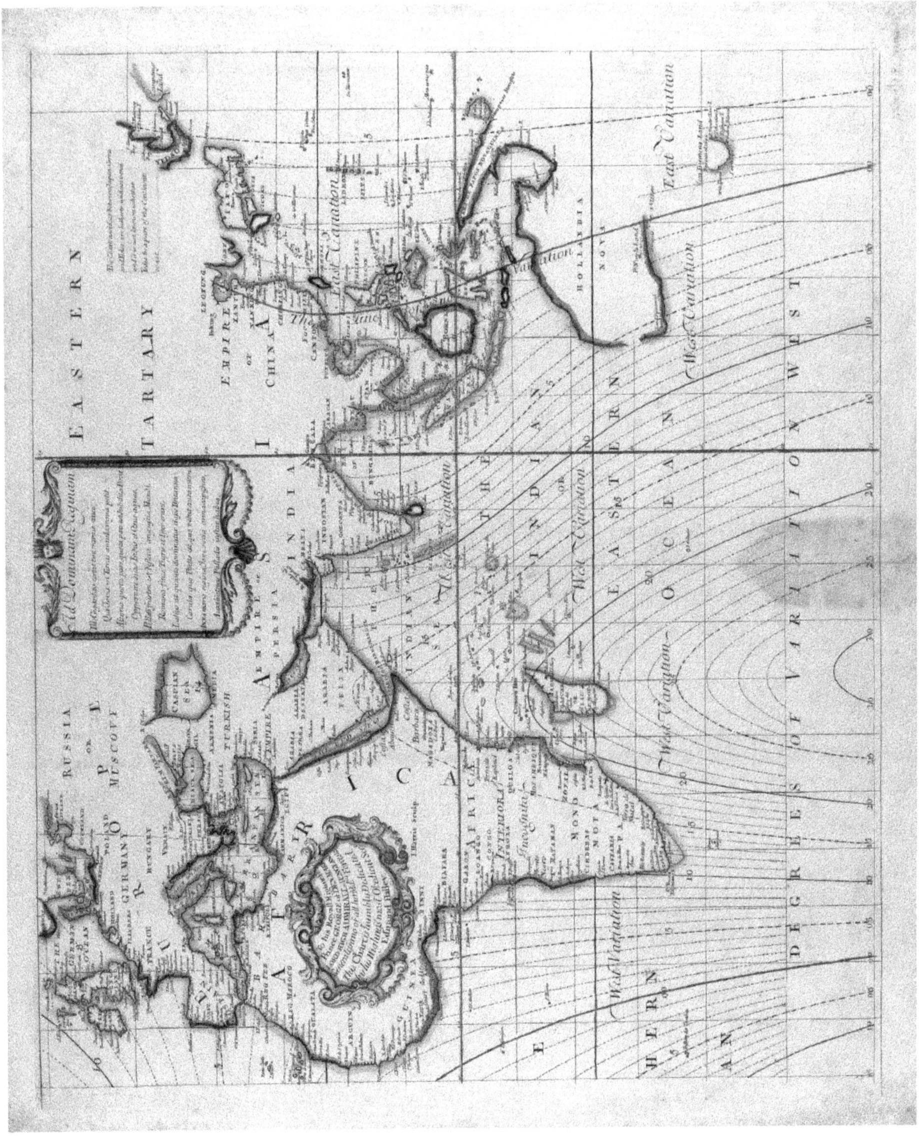
EASTERN
TARTARY
RUSSIA
MUSCOVY
GERMANY
AFRICA
PERSIA
CHINA
East Variation
West Variation
DEGREES OF VARIATION WEST

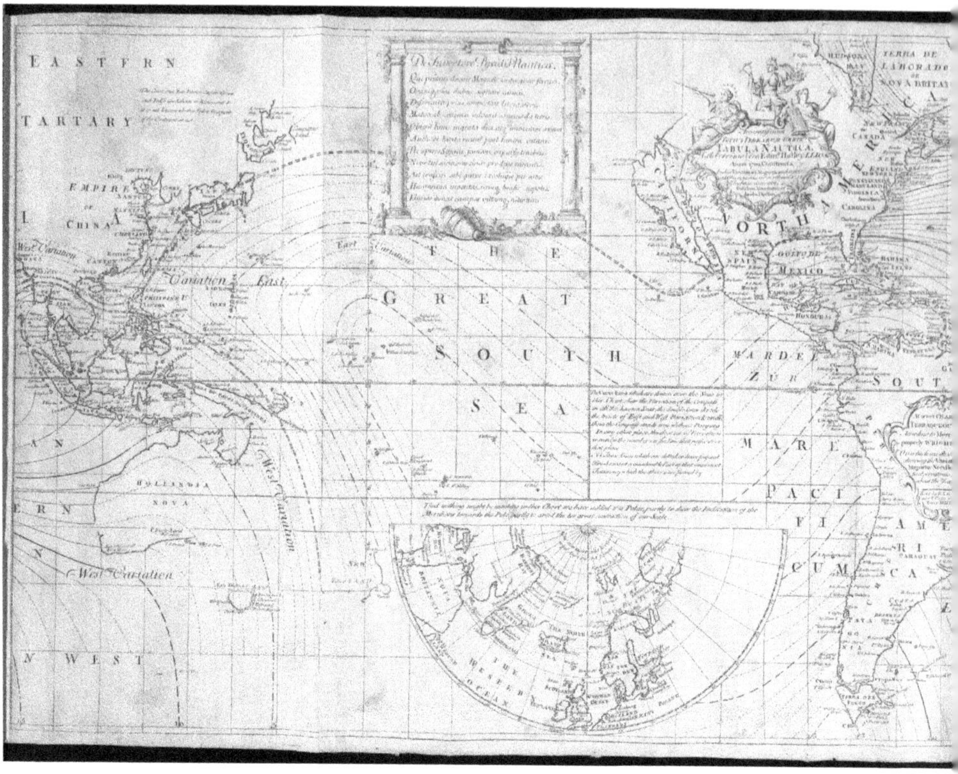

Figure 10: William Mountaine and James Dodson, *A Correct Chart of the Terraqueous Globe according to Mercator's, or more properly Wright's Projection; On which are describ'd Lines, shewing the Variation of the Magnetic Needle, according to observations made about the Year 1744* (1746); engraving, 520 x 1,430mm, Lawrence H. Slaughter Collection, New York Public Library.

goes to Sea'.[79] He recommended that someone be employed to draw a new map and return magnetic variation to its original utility. Halley had noted this drawback himself in his description to the 1701 Atlantic Chart.[80] The problem was, in fact, that magnetic variation lines could not be pinned down any more than longitude ones, the variation shifting continually, requiring new observations and mapping processes.

It was Mountaine and his colleague James Dodson (d.1757) who attempted to answer this question. Dodson was a mathematician and actuary, and later Master of the Mathematical School at Christ's Hospital. In 1746, they

79 Whiston, *Longitude Discovered by the eclipses* (1741), p.xxi.

80 Reproduced in Thrower, *Three Voyages*, p.367.

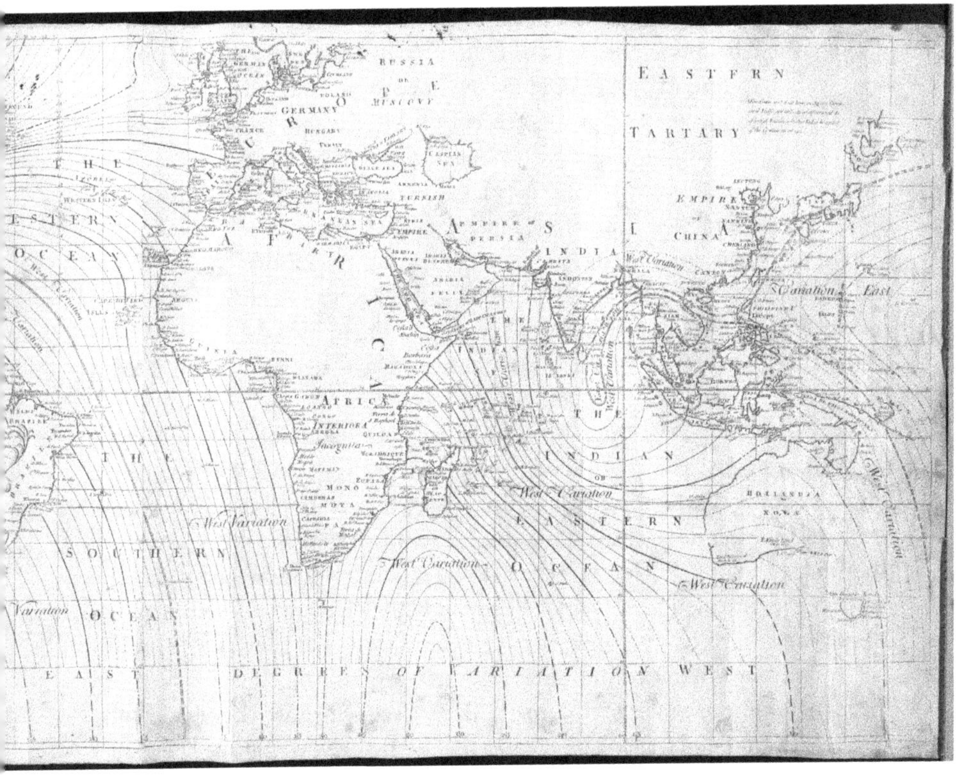

published *An Account of the Methods Used to Describe lines, on Dr. Halley's Chart of the Terraqueous Globe* to accompany and explain the production of their new map. In their introduction, they explained that 'the *Variation Lines* of [Halley's] excellent Performance are now of but small Use, thro' the Length of Time since they were drawn, and therefore the Proprieters of the *Chart* have been at the Expence of substituting others'.[81] They advertised for and themselves undertook a range of observations to correct and extend Halley's lines, outlining how, 'a great Number of Observations made near the same Time, being thus transferr'd to the *Chart*, enabled the Constructors to approve of some and reject others, according as they were supported or not, by concurrent Testimony; and thence to draw Lines representing the Variation at that Time'.[82] It is interesting that, despite the project to update Halley's map, the new lines were, therefore, clearly partly based on their relationship to the old ones, as well as on new observations where Mountaine and Dodson could acquire them. In its design and framing, their map (Figure 10) copies

81 Mountaine, *Account* (1746), p.4.
82 Mountaine, *Account* (1746), p.7.

Halley's, even using many of the same text boxes and cartouches, as well as the same label inscriptions. Like Halley, they explained that where their lines were less sure, they had used dotted ones to make this clear. They also made some important changes, however, joining the two halves of Halley's map, which moved the 'Great South Sea' into the large central space. They also filled it with dotted lines of variation, and added a definitive line dividing east and west, therefore replacing Halley's final self-deprecating sentence on this area of the globe with the comment that 'Those Lines which are dotted, or have frequent Breaks, want a considerable Part of that concurrent Testimony which the others were formed by'. The dotted nature of the lines was used as a means of drawing them but also making them inconcrete.

In the following decade, Mountaine and Dodson sought to re-draw these lines again, and this time enlisted the help of the Royal Society to advertise for accurate observations. In 1753, they wrote to the *Philosophical Transactions* explaining the history of the attempt to map magnetic variation. They made the link to Mercator charts explicit, outlining how 'about the year 1700, the justly celebrated Doctor Edmund Halley having collected together a great number of observations, made on the variation of the needle … drew (on a mercator Chart) certain lines'.[83] Again we see the constant focus on the lines as the essence of the map. They discussed how they had found that the variation changed sufficiently that '(when the writers of this paper endeavoured about the year 1744, to draw thereon, other lines to answer the purposes above mentioned) they found that those laid down by Doctor Halley were grown entirely useless'. Their work had, however, meant that 'they were enabled to draw the proper lines over the most frequented seas', as we have seen.[84] They now argued that, if these lines could be extended over land as well as the seas, they would allow the whole world to be easily mapped. They had received no responses to their advertisements in the newspapers in 1744, so now hoped for help from the Royal Society's networks.[85] The resulting chart (Figure 11) appeared in 1758, with a second accompanying pamphlet of the same title as in 1746.[86] Mountaine and Dodson made an interesting addition to their otherwise largely identical introduction, in which they explained the projection and line placing on the map, saying,

83 Mountaine and Dodson, 'An Attempt to point out, in a concise manner', read 20 March 1755, in *Philosophical Transactions* (1753), p.875.

84 Mountaine and Dodson, 'An Attempt', p.876.

85 Mountaine and Dodson, 'An Attempt', pp.876–8.

86 Mountaine, *Account* (1758) is textually similar, but has been re-set and extended for this pamphlet.

> This *Chart*, upon which the Variation Curves are described, is projected upon Mr. *Wright*'s Principles, and consequently whatever Problems are usually solved upon the said *Chart* (commonly called *Mercator*'s) adapted for that business alone, may be performed also upon this, with equal Facility, except the Determination of the Course, or Bearing of one Place from another, the Rhumb lines not being laid down, as is usual in others for that purpose, on Account of the Confusion that would arise by *these* intersecting the Variation Curves.[87]

They essentially argued for variation lines replacing rhumb lines as a tool for navigation, and for the need for the variation to work with the Mercator grid of latitude and longitude. We can surmise that they learnt this from Halley's Atlantic Chart. Yet their accompanying map retreated from the confidence shown in their first. The lines have noticeably moved west, but some that were complete have now become dotted and the 'Great North Sea' has again become an empty white space. The area of sea east of 'Terra de Labrador de Nova Britannia' has likewise become white, which even Halley had filled with lines in 1702. The magnetic variation lines were becoming increasingly unreliable.

Yet these lines still retained a power and authority throughout Harrison's interaction with the Board. In 1772, John Robertson (1712–76), the mathematician and master of the Royal Naval Academy at Portsmouth, who had been involved in observations for the Jamaica trial of Harrison's watch in 1761, published *The Elements of Navigation*. This new edition included noticeable changes from its predecessor published in 1764, including an entirely new section discussing how 'to find the longitude at sea'.[88] In this, before he discussed the history of Harrison's clocks and of work on the lunar distance method, Robertson suggested a chart of magnetic variation as a possible method, and gave the history of Halley, Mountaine, and Dodson's work. He went on to recommend that 'the engraving of new plates on every publication of such charts, is a charge too great for private persons to incur; and should, like the nautical almanack, be done at the expence of the public'.[89] Thus, Robertson, at least, thought magnetic variation lines were reliable enough to be made a standardised print publication for all naval vessels. In 1775, Samuel Dunn, whom we have already seen defending the Mercator projection against William West's criticisms, produced his own contribution to the magnetic variation solution. In *The Navigator's Guide, to the Oriental or Indian Seas* he focused on an area that had been drawn from

87 Mountaine, *Account* (1758), p.8.
88 This previous edition is Robertson, *Elements of Navigation* (1764).
89 Robertson, *Elements of Navigation* (1772), ii, pp.340–1.

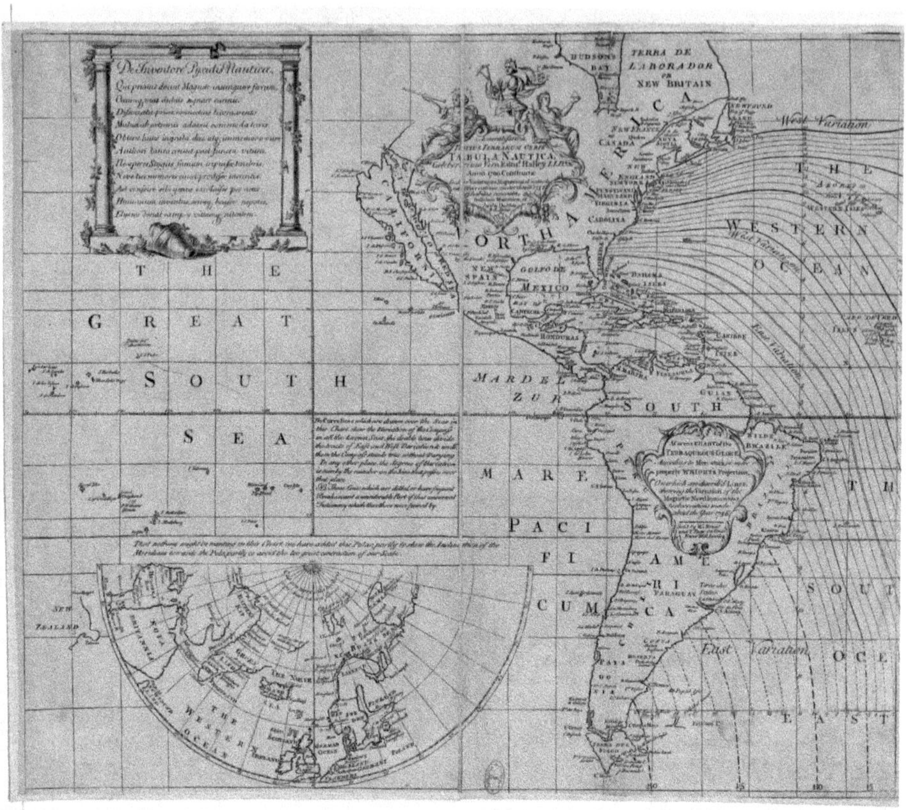

Figure 11: William Mountaine and James Dodson, *A correct Chart of the Terraqueous Globe According to Mercator's or more properly Wright's Projection; On which are describ'd Lines, shewing the Variation of the Magnetic Needle, according to observations made about the Year 1756* (1758); engraving, 520 x 1,220mm, source gallica.bnf.fr / Bibliothèque Nationale de France.

others' observations by Halley, and dedicated it to the East India Company, thus asserting its relevance beyond the Navy. He discussed the history of the work done by Halley, Mountaine, and Dodson but, significantly, criticised 'The smallness of the scale to which all those former charts have been drawn, [which] could not but render them very unfit for any attempt of the longitude at sea … however correctly they may be supposed to have been drawn'.[90] Scale is not a factor that had been discussed before, and we note the prominent circular scale of both degrees and rhumbs that now fills

90 Dunn, *Navigator's guide*, p.10.

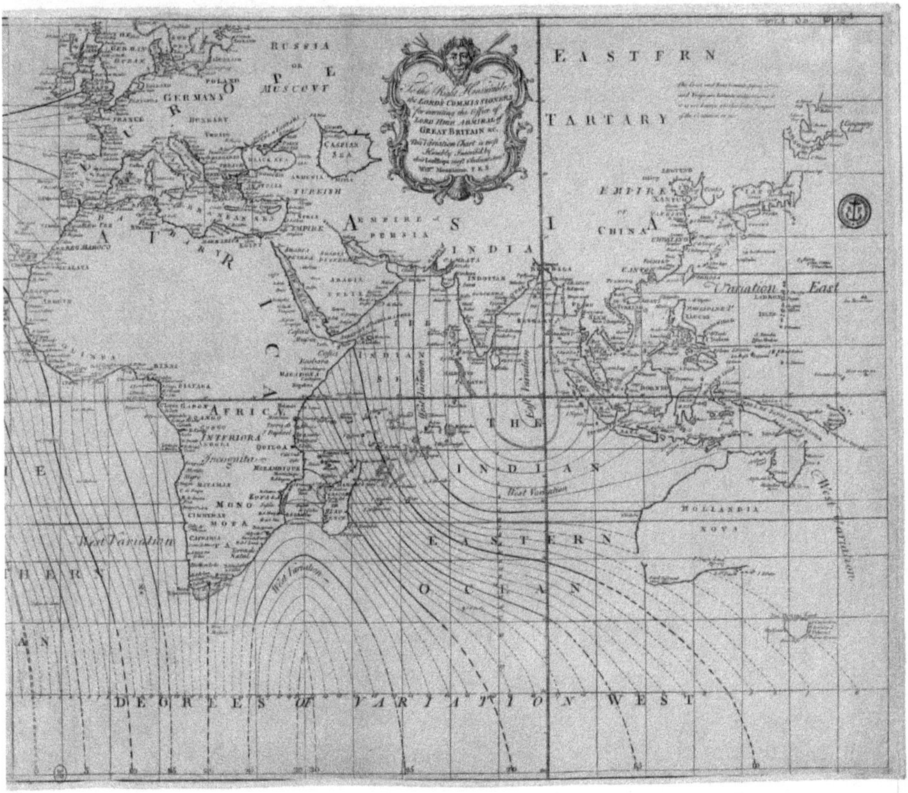

Europe and Africa in Dunn's accompanying chart (Figure 12). He went on to emphasise his own improvements to the chart, saying, 'I have been the more particular in this case, to shew what errors might have happened by the introduction of more sea coasts, where the longitudes have not been well ascertained'.[91] In directly instructing the reader how to read the chart along the different sets of lines, he nicely epitomised what all our protagonists had argued: 'having the latitude of the place somewhat near the truth … guide your eye along that latitude, on the Variation Chart … until you come to such a curve line of the variation as you have observed'.[92] He showed how the curved variation and straight latitude lines could work together.

91 Dunn, *Navigator's guide*, p.13.
92 Dunn, *Navigator's guide*, p.16.

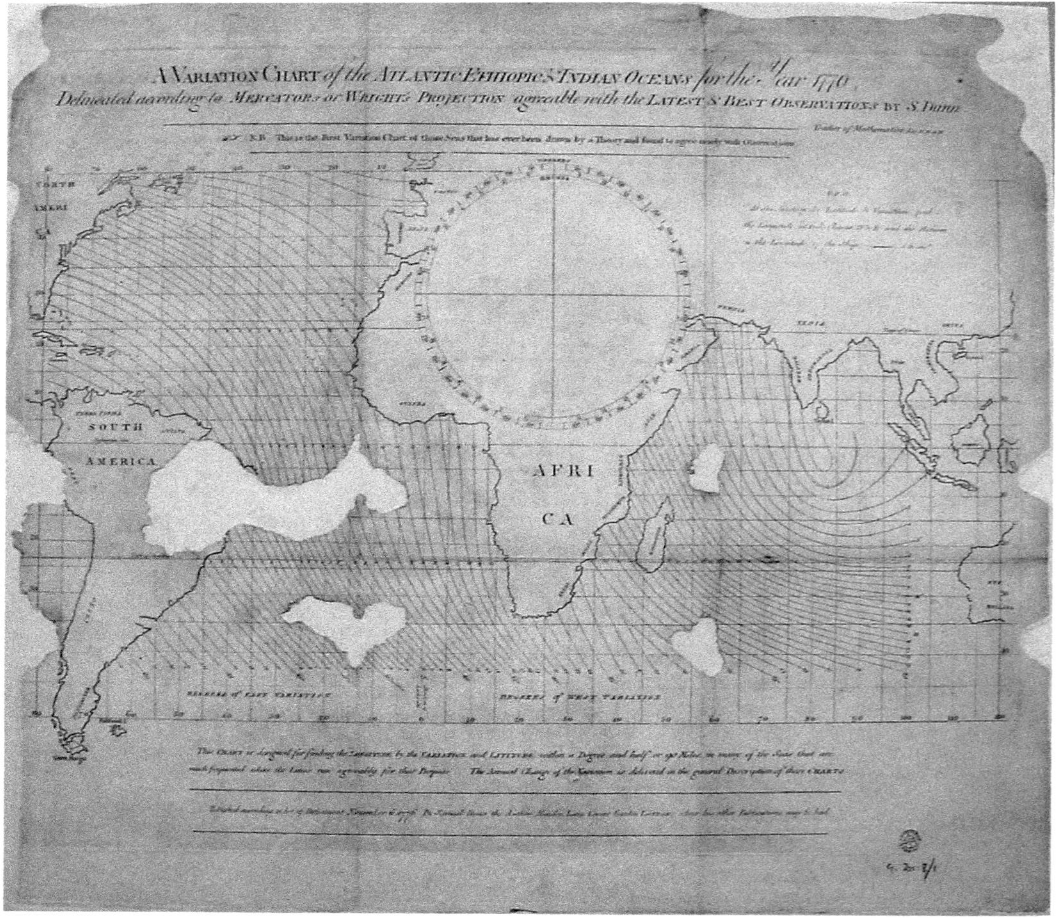

Figure 12: Samuel Dunn, *A variation chart of the Atlantic, Ethiopic and Indian Oceans for the year 1770, delineated according to Mercator's or Wright's projection agreeable with the latest and best observations by S. Dunn* (1776); engraving, 630 x 510mm, Royal Museums Greenwich, London.

'the Shores, which are Places of the greatest Danger'

Dunn was not the only one concerned with the accuracy of coastal longitudes, however. Longitude lines, or magnetic variation lines, are, after all, only reliable if they join up a series of accurate points.[93] We have seen already how William Whiston commented on the accuracy of Halley's map of magnetic variation. It was, in fact, the accuracy of coastal mapping that really interested him, and which marks the only direct interaction of the Commissioners of Longitude with this early story of magnetic mapping, and the only cartographic proposal to be funded by them.[94] Whiston's proposal was to use magnetic dip, rather than variation, plotting how far the compass needle varied from the horizontal in pointing north, but to plot this on a map in the same manner as Halley. He published discussions of this in 1719 and 1721, both entitled *The Longitude and Latitude Found by the Inclinatory or Dipping Needle*, where the second materially added to the explanations and figures of the first. The 1719 version included one chart of the east coast of England and the English Channel (Figure 13), to make Whiston's point that lines of magnetic dip could be used in conjunction with the grid of latitude and longitude, 'as the View of my Map hereto belonging, which is made from the best Observations, will readily demonstrate'.[95] Alongside this he included two small diagrams demonstrating how the dip could be observed and then plotted onto the map. He asserted that charts could easily be drawn by joining current observations into suggested lines, which pilots could then correct as they sailed, and gave an example using the pictured map.[96] This would in particular, Whiston contended, allow coastal areas of Britain to be plotted accurately, and especially the Scilly Isles, which were periodically discussed as a particular danger for British navigation. He proposed this as the first activity for the Commissioners to fund.[97]

The 1721 edition expanded this proposal with a further chart and diagrams (Figure 14) showing Whiston's method, as well as a detailed 'historical preface' explaining the history of magnetic variation and mapping. Notably, it was John Senex who published this edition. In the preface, Whiston explained how his idea of mapping had been inspired by Halley's work, saying,

93 Tim Ingold has discussed points or dots on maps as comparable to knots, into which lines are wound and tied; Ingold, *Lines*, pp.73–4.

94 The Commissioners would receive many proposals for using magnetic variation in the 1780s–1820s. See RGO 14/42 and 14/43.

95 Whiston, *Longitude and latitude found* (1719), p.23.

96 Whiston, *Longitude and latitude found* (1719), p.28.

97 Whiston, *Longitude and latitude found* (1719), pp.30–1.

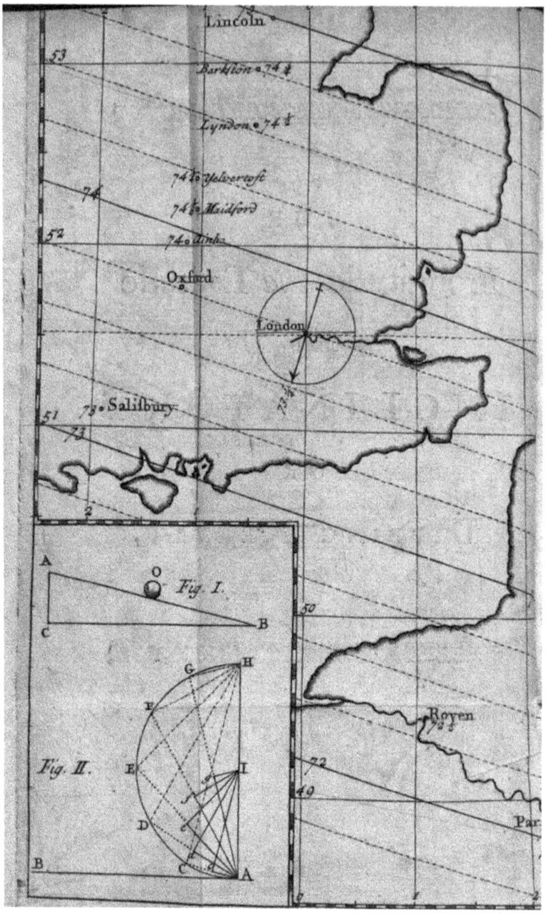

Figure 13: Plate from William Whiston, *The longitude and latitude found by the inclinatory or dipping needle* (London, 1719); engraving, 186 x 110mm, Royal Museums Greenwich, London.

> by such an inartificial way of describing the Curve Lines of equal Variation, how irregular soever, upon a Map, he had already given some Help for the Discovery of the Longitude … I considered, that the Lines of equal Dip could hardly be more irregular than those of the Variation; and well knew their Mutation was a great deal slower.[98]

Alongside this, Whiston now included an additional chart of a slightly larger British coastline, which included the west coast around Liverpool. This showed the magnetic dip in more detail, with more lines, more marked towns, and a table of their exact variations, as well as a figure explaining how to use these to calculate longitude. He commented that he hoped no one would begrudge him the longitude reward.[99] It was not until 1741, however, that he managed to

98 Whiston, *Longitude and latitude found* (1721), pp.xxvi–vii.
99 Whiston, *Longitude and latitude found* (1721), p.xviii.

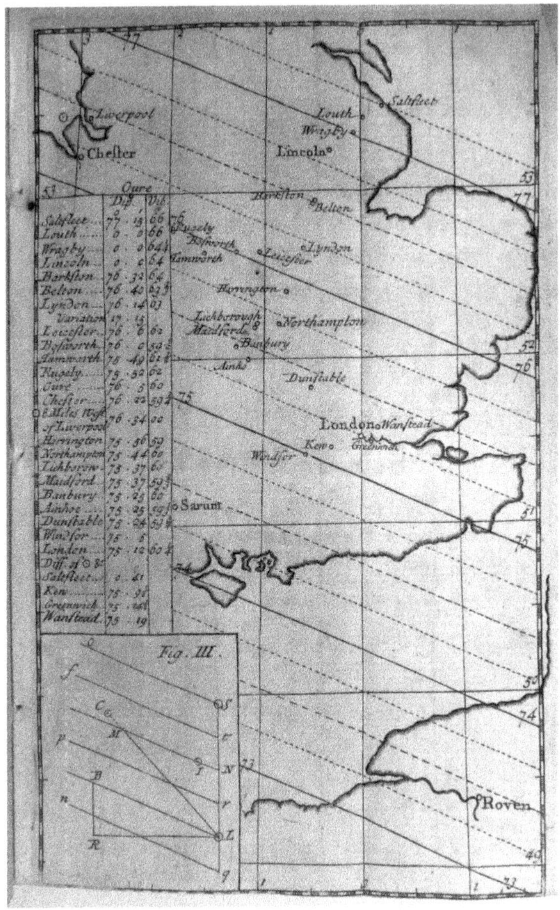

Figure 14: Plate II from William Whiston, *The longitude and latitude found by the inclinatory or dipping needle; wherein the laws of magnetism are also discover'd* (London, 1721); engraving, 188 x 115mm, Royal Museums Greenwich, London.

present his proposal to the Commissioners. At a meeting on 16 January of that year, he represented to the Board of Longitude how effective his observations had proved in charting the British coast, but that he could not afford to extend them further, and therefore requested five hundred pounds assistance.[100] He was awarded this, at the same time that Harrison was awarded the equivalent sum to continue work on his timekeeper, as Richard Rolt observed in his *New Dictionary*. What is interesting is Whiston's use here of a clause in the 1714 Act to which attention does not seem to have been paid at any other time. This allowed for half of a reward to be paid when a solution was shown to improve 'the Security of Ships within eighty Geographical Miles of the Shores, which are Places of the greatest Danger'.[101] Whiston also claimed to

100 Meeting of the Commissioners, 16 January 1741, RGO 14/5: 6.

101 RGO 14/1: 12r.

have been directly involved in the addition of this clause to the Act during its passage through Parliament in 1714.[102] Whiston was, it seems, much better at utilising the definitions of this ambiguous act than Harrison.

Harrison was, however, also concerned with the problem of accurately locating points on land in order to compare them to the longitude at sea. He discovered how crucial this was during the process of trialling his watch for the Board of Longitude. On William's return from the first trial to Jamaica, among other concerns, the Commissioners of Longitude decided that they could not be certain of the watch's accuracy 'for we do not know the Longitude of Port Royal'.[103] After much wrangling, both at the Board and in Parliament, it was decided that William would take the watch on a second trial to Barbados. Harrison recorded a lengthy discussion with the Astronomer Royal Nathaniel Bliss (1700–64), over the destination of the trial, where Antigua was rejected for its longitude being accurately known, and Barbados chosen for the opposite reason. Harrison concluded from this complex state of affairs that 'my fate is very hard that I must not now go to a place because you say the Longitude of it is known, whereas a little time ago you could not give me the Reward because you said the Longitude of that place was not known'.[104] We can see the problem for the Board in needing to spread its resources as widely as possible, thus in wanting to use trials to establish the longitude of key ports in the West Indies, while also needing an accurate comparison for the longitude at sea found using Harrison's watch.

Jonathan Swift satirised precisely this problem with mapping in *Gulliver's Travels*. In the novel, Swift reproduces maps of the lands that Gulliver visits, which float in a field of unmarked white sea, where a small ship sails at the mercy of the blank page and a couple of sea monsters (Figure 15). Yet each new land discovered is announced in the text with the nearest known latitude and longitude coordinates, stated with complete confidence as if uncontested. Balnibarbi, for instance, we are told is near to the 'Latitude of 46 N. and of Longitude 183' according to Gulliver's own observation. Yet, as we will see in Chapter 3, Swift was extremely dubious about the project of finding longitude at sea. He would be particularly amused at how these coordinates are used by modern editors to locate Balnibarbi on maps.[105]

---

102 It was in relation to Whiston and Ditton's projectile scheme that the split reward was discussed. *Journal of the House of Commons* (11 June 1714), pp.677–8. Whiston claimed responsibility for this addition in *Longitude Discovered by the eclipses* (1741), p.vii.

103 *HJ*, p.31. The full record of these decisions can be read in RGO 14/5.

104 *HJ*, p.63.

105 Swift, *Gulliver's Travels*, p.143. Rogers notes that this places Brobdingnag in Southern Waters, making a link to the South Sea Scheme discussed in Chapter 3, see 'Gulliver and the Engineers' in *Eighteenth Century Encounters*, pp.11–12. These coordinates

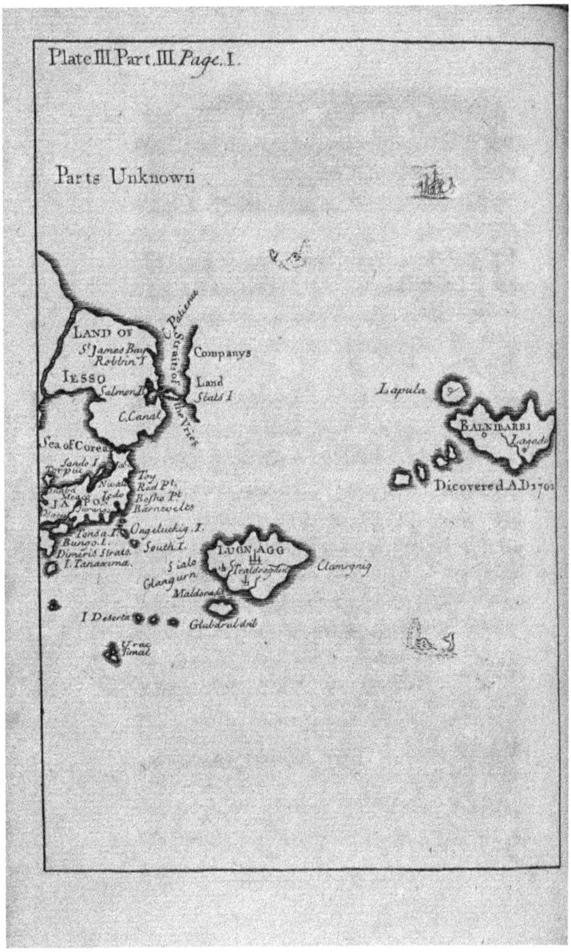

Figure 15: Plate III, Part III from Jonathan Swift, *Travels into several Remote Nations of the World. In four parts. By Lemuel Gulliver, first a surgeon, and then a captain of several ships* (London, 1726); engraving, 196 x 116mm, reproduced by kind permission of the Syndics of Cambridge University Library [Syn.7.72.14].

What Swift did with the floating island of Laputa was to make a joke of just the cartographic problems and arguments that we have been discussing. Laputa is a physically floating island, powered by an extremely powerful magnet which allows the inhabitants to fly it, like a sailing ship, over the subjugated land of Balnibarbi, and use it as a tool of oppression.[106] In a smaller-scale chart of the kingdom (Figure 16), Swift included a diagram of how turning the loadstone allowed the island to be moved, aping precisely the maps and diagrams of men like Whiston.

also locate Brobdingnag east of Japan, an exotic society notoriously closed to most European trade in the period.

106 The wider significance of this as a magnetic scheme is discussed by Lynall in *Swift and Science*, pp.109–19.

Figure 16: Plate IV, Part III from Jonathan Swift, *Travels into several Remote Nations of the World. In four parts. By Lemuel Gulliver, first a surgeon, and then a captain of several ships* (London, 1726); engraving, 196 x 114mm, reproduced by kind permission of the Syndics of Cambridge University Library [Syn.7.72.14].

'EVERY THING THAT MADE THAT PAPER VALUABLE'

THE HARD WORK OF ENGRAVING also led Hogarth to concern himself with acts and definitions. During the process of engraving *A Rake's Progress*, he petitioned Parliament with six other artists and engravers for the passage of the Engravers' Act.[107] This argued for engravers to be granted rights in their designed images in the same way as rights had been vested in written texts by the 1710 Copyright Act. In *The Case of Designers, Engravers, Etchers, &c* that Hogarth and his compatriots presented to Parliament in February 1735, they outlined how 'he who can only follow with his Tool a Line already

107 Act 8 Geo.2 c.13, Parliamentary Statutes online.

drawn, can by colouring the Back of a Print, and placing it properly on a Plate … trace out and delineate the Out-lines, and every remarkable Feature, nay each individual Line of the Print'.[108] We note how they emphasised the transfer of lines onto the page. They argued, further, that this should be considered in the same vein as forgery, asking 'Does not [the copyist] as much injure the [author], as if he counterfeited a Note of his Hand? … He does not indeed steal the very Paper … but he steals from him every Thing that made that Paper valuable'.[109] It is particularly interesting that the witnesses called to comment on the proposal to the House of Commons showed originals and pirate versions to make their point.[110] The resulting Engravers' Act copied precisely the model of the earlier Copyright Act, establishing a fourteen-year copyright period for new designs.[111] Engravings would now have to include the name of the proprietor and the date of the first copy.[112] The publicity surrounding it served to remind the book publishers of the recent lapse in their rights, and a new copyright bill was introduced to Parliament in the same session.[113] Map publishers were particularly quick to make use of the Act to protect new projections and charts based on new observations.[114] Hogarth had helped to establish the notion of ownership in lines on the page, and it became colloquially known as the Hogarth Act. He became synonymous with these notions of copyright, as Harrison did with the definition of longitude. In the next chapter, we will see how these notions of copyright became further bound up with the difference between an instrument and an image of an instrument on paper, and how Harrison again became the focus of controversy.

108 Hogarth, *Case*, p.9.
109 Hogarth, *Case*, p.9.
110 *Journal of the House of Commons* (14 February 1735), p.381.
111 Feather, 'Publishers and the Pirates', pp.7–8.
112 Hunter, 'Copyright Protection', p.136.
113 Feather, 'Publishers and the Pirates', pp.8–11.
114 Hunter, 'Copyright Protection', p.146.

Figure 17: Johann Sebastian Müller, 'No. 14 Vol 1 Part 2 p.147' from *The works of Jonathan Swift, D.D. Dean of St. Patrick's, Dublin* (London, 1755); etching, 162 x 109mm, © Trustees of the British Museum.

2

Paper Instruments

WHILE SWIFT USED THE FLOATING ISLAND of Laputa to make a joke about maps and diagrams in *Gulliver's Travels*, he made the inhabitants of his island critique a different kind of visual activity. Like Hogarth's longitude lunatic, the Laputans cover the walls of their court with illustrations of instruments, both musical and scientific, in Johann Sebastian Müller's engraving for the novel (Figure 17). Such images were crucial to many of the pamphlets that proposed new instruments to solve longitude, representing the instruments or explaining their development or operation. They also became crucial to Harrison's discussions with the Board of Longitude. These images involved a different type of line combination to maps, one that required shading and perspective, still the representation of a three-dimensional shape on a flat plane, but the visualising of its shape rather than flattening of its surface like a map. They raised a whole different set of questions about visual truth on paper.

We remember that John Harris announced in the preface to his *Lexicon Technicum* that he wanted to make a dictionary of things as well as words. He therefore included detailed engravings of the instruments that he discussed, from air pumps to microscopes, a practice that was also copied by Chambers in the *Cyclopædia*.[1] If we look back at Halley's magnetic variation maps, our eye is caught by the three muses who recline along the top of the title cartouche, and who remained in place in all Mountaine and Dodson's re-drawings

1 Bradshaw, 'Ephraim Chambers's *Cyclopaedia*' in McDermott, *Ashgate Critical Essays*, pp.250–1.

(Figures 8–11). They are in the process of using instruments: a telescope, ring dial, backstaff, mariner's compass, pair of compasses, and prism, along with a ship to which to apply them, and a sheet of paper on which to inscribe their results. All such instruments were involved in attempts to find longitude at sea, and all required improvement in the eighteenth century to become sufficiently accurate. Like our mapping protagonists, instrument proposers commented on and adapted their images as part of asserting the efficacy of their scheme.[2] Timekeepers and quadrants became particularly central to such discussions.

Eighteenth-century illustrations of astronomical instruments can be seen within a dual framework of the 'professional' and 'amateur' illustration. The 'professional' illustration, or 'portrait of an individual object', was to be judged by the maker's scientific peers for its accuracy, while the 'amateur' illustration of the generalised instrument was to be bought by the public.[3] On the boundary between 'professional' and 'amateur' illustrations sat the illustrations that accompanied accounts of instruments published in their own right, not linked to measurements.[4] Here, the illustration became as much a technical drawing as a portrait. Illustrations allowed proposed instruments to become 'impartial' by allowing readers to view and judge for themselves, like art history illustrations, but they also worked to create an impression of the reality of an instrument as something that could be invested in.[5] For readers in the eighteenth century such images could become the object in a manner particularly useful for longitude inventions, where the image often pre-dated the instrument's construction.

'All the Meters are not worthy to be compared with my Chronometer'[6]

When Flamsteed sent pamphlet proposals and comments to Abraham Sharp in 1714, he discussed specifically how 'Seignr. Alimari has sent me a Description of his Instrument, with a Figure of it, which was needless, for it was easy to apprehend without the figure, that it was one of the worst

2 Burke, *Social History*, p.100. Daston and Galison discuss how this would become 'objectivity' by the early nineteenth century in *Objectivity*. Lefèvre has considered how early-modern machines between 1400 and 1700 were 'pictured' in Lefèvre, *Picturing Machines*.

3 Bennett, 'Instruments and Illustrations' in Shea, *Science and the Visual Image*, pp.137–54.

4 Bennett, 'Instruments and Illustrations', pp.145–6.

5 Smiles, *Eye Witness*, pp.3, 19.

6 Thacker, *Longitudes examin'd*, p.23.

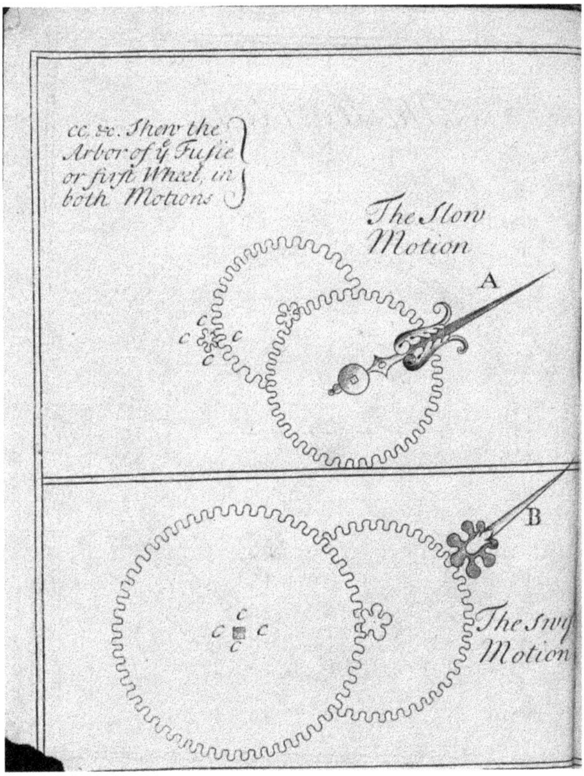

Figure 18: 'The Slow and Swift Motion' from William Hobbs, *A new discovery for finding the longitude* (London, 1714); engraving, 218 x 161mm, reproduced by kind permission of the Syndics of Cambridge University Library [White.c.3].

contrivances for taking the height of the Sun or Stars that was ever thought of'.[7] His irritated manner came, as we have seen, from the sheer number of proposals that appeared in that first year, many of which included expensively engraved figures. What is remarkable in these is both the problematic relationship between illustration and text, and the extent to which such pamphlets engaged with each other in both media. Also in 1714, William Hobbs (1658–fl.1716) published *A New Discovery for Finding the Longitude*, in which he proposed a new spring movement to keep time accurately at sea, rather than a pendulum.[8] He included two plates of simple line-drawn diagrams of parts of his clock. One showed, from the back, the wheels controlling the clock hands, in which one created a slow and one a fast motion (Figure 18). The other showed the 'horologe' face with the hour scale

7 Flamsteed, *Correspondence*, Letter 1,366, p.712. Flamsteed is discussing Alimari's *New method*.

8 Hobbs, *New discovery* (1714). Wigelsworth has concluded that this is not the same Hobbs who wrote *The Earth Generated and Anatomized* that Roy Porter edited in 1981. See Wigelsworth, *Selling Science*, p.128.

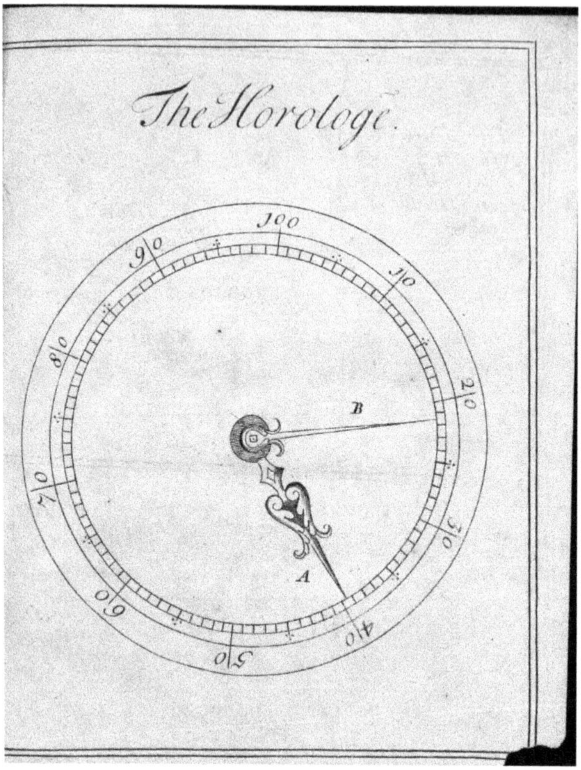

Figure 19: 'The Horologe' from William Hobbs, *A new discovery for finding the longitude* (London, 1714); engraving, 215 x 161mm, reproduced by kind permission of the Syndics of Cambridge University Library [White.c.3].

and two labelled indicators, one for home and one for local time (Figure 19). Nowhere was Hobbs' new spring mechanism illustrated, and nowhere did he refer directly to his illustrations, despite annotating them with key letters. The wheels, likewise, show little technical information, yet the hour scale on the clock face is carefully drawn and the indicators are given decorative moulding and shaded depth. The reality appears to be in the recognisable part of the clock—the face—rather than in the workings. Later in 1714, the otherwise largely untraceable Wiltshireman James Clarke published another horological proposal: *An Essay, Wherein a Method is Humbly Propos'd for Measuring Equal Time with the Utmost Exactness.*[9] In this he proposed keeping time through pouring mercury between a number of different glass vessels, to act essentially like a more exact hourglass. His proposal was accompanied by a simple line engraving (Figure 20) of the whole piece of equipment, on a single page with no title. Although the figure is followed by a half-page 'explanation', this combines poorly with the inadequate labelling. Each gives

9 Clarke, *An essay*. The only other existing record of Clarke's life appears to be his will of 16 February 1721 (National Archives, PROB 11/582/14).

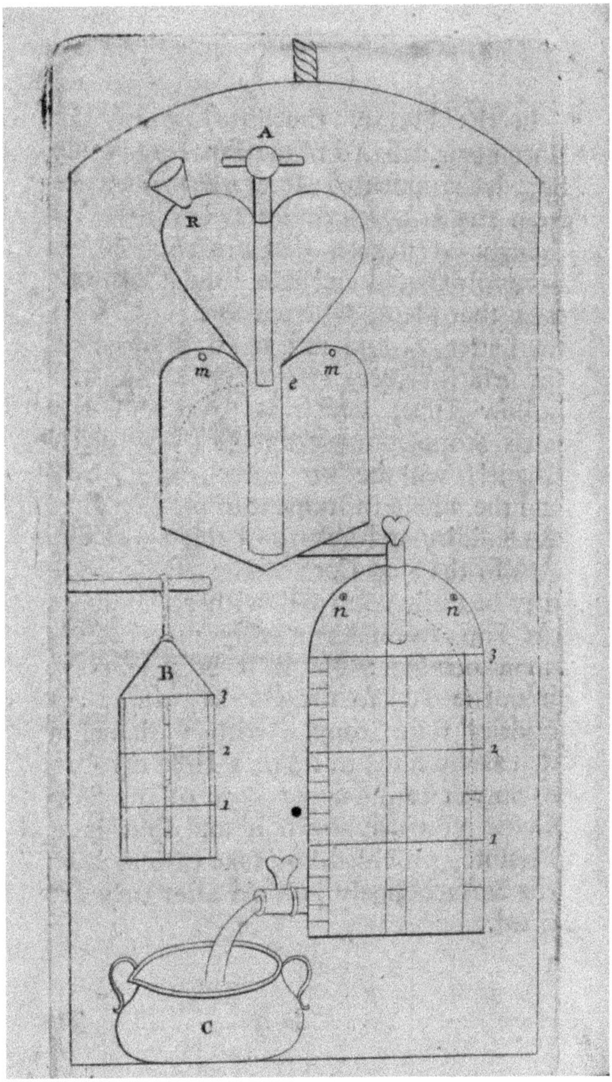

Figure 20: Plate from James Clarke, *An essay, wherein a method is humbly propos'd for measuring equal time with the utmost exactness* (London, 1714); engraving, 168 x 100mm, reproduced by kind permission of the Syndics of Cambridge University Library [IV.25.64(1)].

little understanding of the spatial relationship between the parts of the equipment, of how these parts would work, of how the whole piece would be fixed in the casing, or of how exactly the mercury progresses through the equipment. It is a useful visual reference, especially for the shapes of the flasks that he proposes, but otherwise the illustration adds little credibility to the instrument that Clarke was proposing. It is unclear why he would go to the additional expense of a printed plate that does such little work. The key lettering, at most, works to create an image that is visually 'scientific' in the same manner as Swift's diagram of the island of Laputa.

Jeremy Thacker commented critically on both Hobbs' and Clarke's proposals, among many others, in his own contribution, also in 1714, *The Longitudes Examin'd*, which proposed a more traditional clockwork solution.[10] Thacker is noteworthy for outlining with great clarity and perception the problems involved in making a clock work at sea, including controlling the pendulum swing and allowing for temperature change, as well as employing one of the first uses of the word 'chronometer' to describe a marine timekeeper.[11] His proposal was to put his clock in a vacuum as a means of isolating these problems. He was clearly aware of the particular importance of visual elements in a scheme. He commented of Alimari's scheme, almost as though he was responding to Flamsteed, 'let not our *Italian* Gentleman be discourag'd from giving us the Figure of his Engine, for he is not the first Professor of Mathematicks who has fallen short of the Longitude'.[12] In the postscript, he added comments on two proposals that had been published after the printing of the main text, one of which, by Francis Haldanby (fl.1714) proposed a chart to be used as an instrument.[13] Thacker commented satirically on

> the Dignity of the Author (who has put his Name in the Copper Plate to show the whole Invention to be his own) … His work then consists of a Copper Plate, grav'd by *John Senex* … with a short *Apology* for his *Brevity*, which is *very good* … because it is so easily understood; the whole matter being known to every common Seaman.[14]

We see the appearance of John Senex again in these comments, as well as Thacker's awareness of mapmaking and navigational practices. All types of engraved line seem to have interested him.

Thacker paid particular attention to the proposals by Hobbs and Clarke. He criticised the lack of clarity in Hobbs' explanation of how his timekeeper

10 Thacker's status as a 'serious' contributor to the longitude problem has been much debated since Pat Rogers proposed this pamphlet as a Scriblerian hoax in 'Longitude forged'. This is discussed in Chapter 3.

11 For Thacker's credentials as a serious clockmaker see Jonathan Betts and Andrew King's response to Rogers. Thacker's pamphlet was also mentioned by the Board of Longitude at a meeting in March 1809. The Commissioners questioned the clockmaker Joseph Manton on his proposal, asking 'are you aware that the idea of keeping the Watch in Vacuo is not new, but was published in the year 1714 by Jeremiah Thacker of Beverley in Yorkshire[?]' (Meeting of 2 March 1809, RGO 14/7: 125–6).

12 Thacker, *Longitudes examin'd*, p.5. This jibe is also partly aimed at Whiston.

13 Thacker, *Longitudes examin'd*, pp.21–3. The two proposals discussed are Coster, *Practicable Method*, and Haldanby, *Attempt to discover*.

14 Thacker, *Longitudes examin'd*, pp.21–2.

could function at sea, saying drily, 'Before Mr. H[obb]s thinks of sending his Spring-Movement to Sea, let him know how to make the month of *June* in one Year, just as hot as the same month in another … and then he may polish his Pivots, and make an Aetherial Oil that won't thicken and increase the Friction of his Watch'.[15] He implied, essentially, that the crucial mechanisms of Hobbs' design were insufficiently thought through, compared to the superficial concerns of oil and clock faces. Clarke's scheme he wrongly attributed to 'J. H.' but discussed at some length, complaining that he 'has such a fertile Brain at Invention, that I know not how to deal with him … if I demolish one of his Movements, he'll pop upon me with another'.[16] He commented that the mercury would not remain constant in different atmospheres and that Clarke's suggestion of a pendulum kept constant by being in the centre of a ship 'will not move his Watch-Work. If he would have this Instrument go, let him consult about it with the Inventors of a *Perpetual Motion*'.[17] Thacker's illustration was also markedly different to both Hobbs' and Clarke's (Figure 21). He included a large foldout plate with four figures, carefully shaded and labelled, with an extensive explanation in the text. They not only illustrated his invention in full, but also illustrated his discussion of pendulums, his crucial experiment on springs, and his key method of observation using his neighbour's chimney.[18] He uses the single page to relate different pieces of data between his different images, and back into the explanatory text.[19] All four figures work together to give a clear perception of Thacker's active progress to his solution. They give a physical reality to his proposal which Hobbs' and Clarke's lack, and therefore make his criticisms likewise seem more weighty.

What is striking is that these criticisms appear to have had a visible effect on both Hobbs and Clarke, causing them to re-write and re-illustrate their pamphlets. Hobbs produced an extended second edition of his pamphlet in 1716, in which he added to the title page a description of himself as 'Philo Mathem', a lover of the mathematics, echoing Thacker's title self-description as 'Philomath, Well-wisher to the Twenty Thousand Pounds'.[20] In the text,

15 Thacker, *Longitudes examin'd*, p.3.

16 Thacker, *Longitudes examin'd*, p.5.

17 Thacker, *Longitudes examin'd*, pp.6–7. This references ideas of projecting discussed in Chapter 3.

18 Rogers discusses the relationship between Thacker's images and the 'scientific' images used by Swift in *Gulliver's Travels* in 'Longitude forged'.

19 This idea of diagram has been discussed in Bender and Marrinan, *Culture of Diagram* where they particularly relate it to practices seen in works like Chambers' *Cyclopædia* and Diderot and D'Alembert's *Encyclopédie*.

20 Hobbs, *New discovery* (1716); Thacker, *Longitudes examin'd*, p.iii.

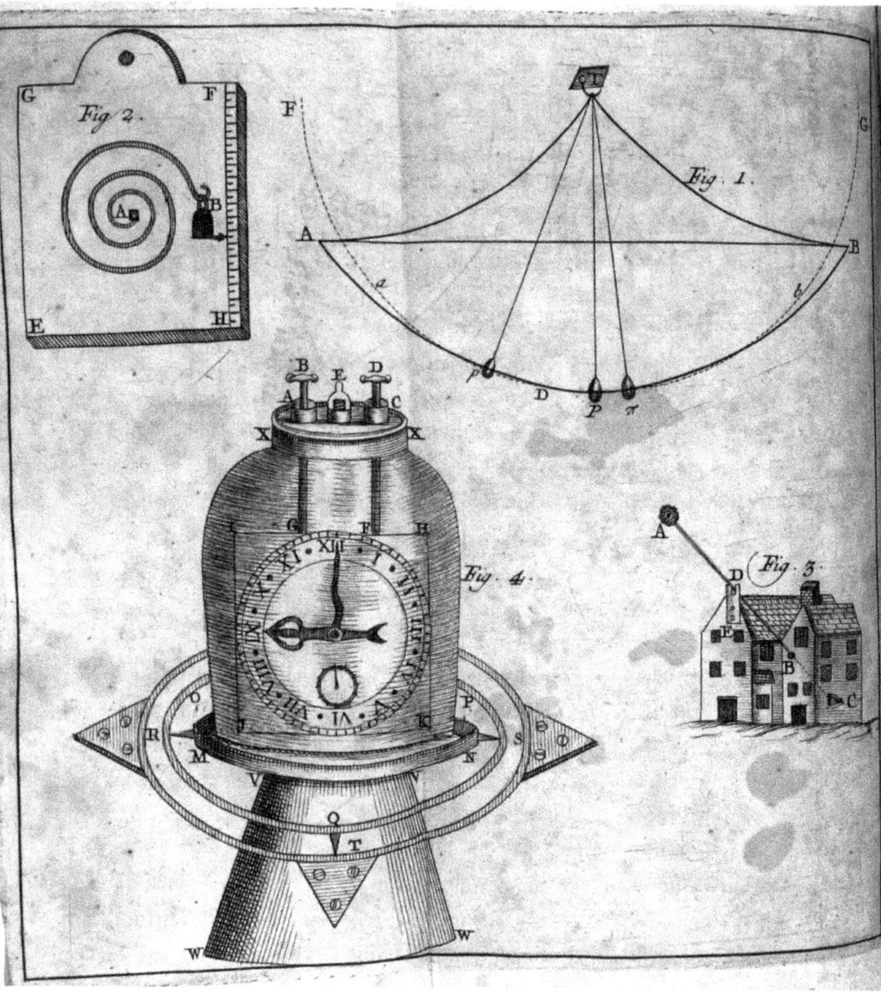

Figure 21: Plate from Jeremy Thacker, *The longitudes examin'd* (London, 1714); engraving, 193 x 164mm, by permission of the Master and Fellows of St John's College, Cambridge.

he added more extensive mathematical workings as well as outlining a possible trial, including a description of how to note down the rate of going in order to allow for variations caused by climate.[21] He also had the plate for his second image re-engraved to emphasise that the clock could be used in all locations by applying its different rates of going. The title to this image now reads 'The Horologe Universal: or The Variable and Infallible Horologe'

21 Hobbs, *New discovery* (1716), pp.3, 5–8.

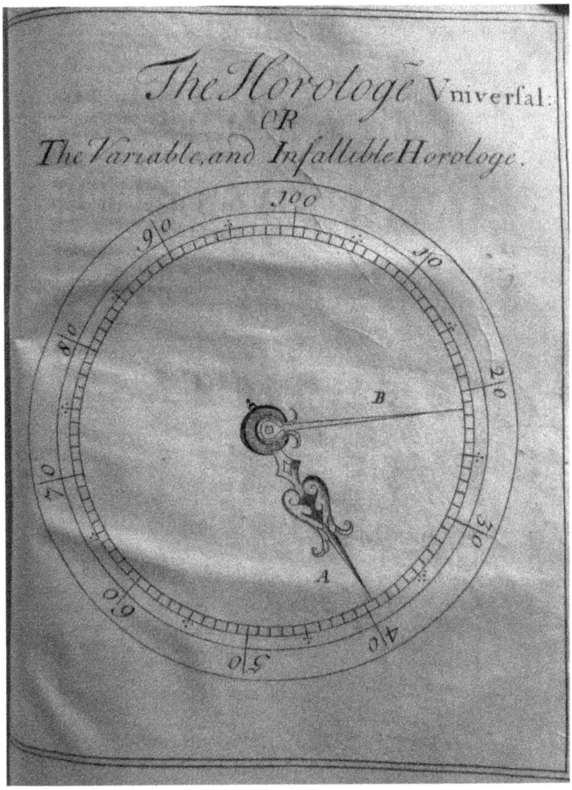

Figure 22: 'The Horologe Universal: or The Variable and Infallible Horologe' from William Hobbs, *A new discovery for finding the longitude* (London, 1716); engraving, 210 x 158mm, by permission of the President and Fellows of Magdalen College, Oxford.

(Figure 22). More strikingly, Clarke published an entirely new pamphlet in 1715—*The Mercurial Chronometer Improv'd*—which vastly expanded and explained his original proposal. He responded directly to Thacker in the text, saying, 'how far a certain merry *Longitudinarian's* Watch, shut up *in Vacuo*, is likely to succeed, I must leave to better judges than myself'.[22] Most effectively, his illustration completely changed to function like Thacker's (Figure 23). He now included a large foldout plate with six carefully shaded illustrations of instruments in perspective projection, which were extensively explained in the text. He referred specifically back to the diagram in the original *Essay*, to explain how he had improved his design.[23] The plate also now included a diagram to show 'the whole operation of sealing the Glasses'[24] (for the mercury) which featured the person performing this task and the candles involved. He followed Thacker's lead in building up a picture of the

22 Clarke, *Mercurial chronometer*, p.4.
23 Clarke, *Mercurial chronometer*, p.8.
24 Clarke, *Mercurial chronometer*, p.27.

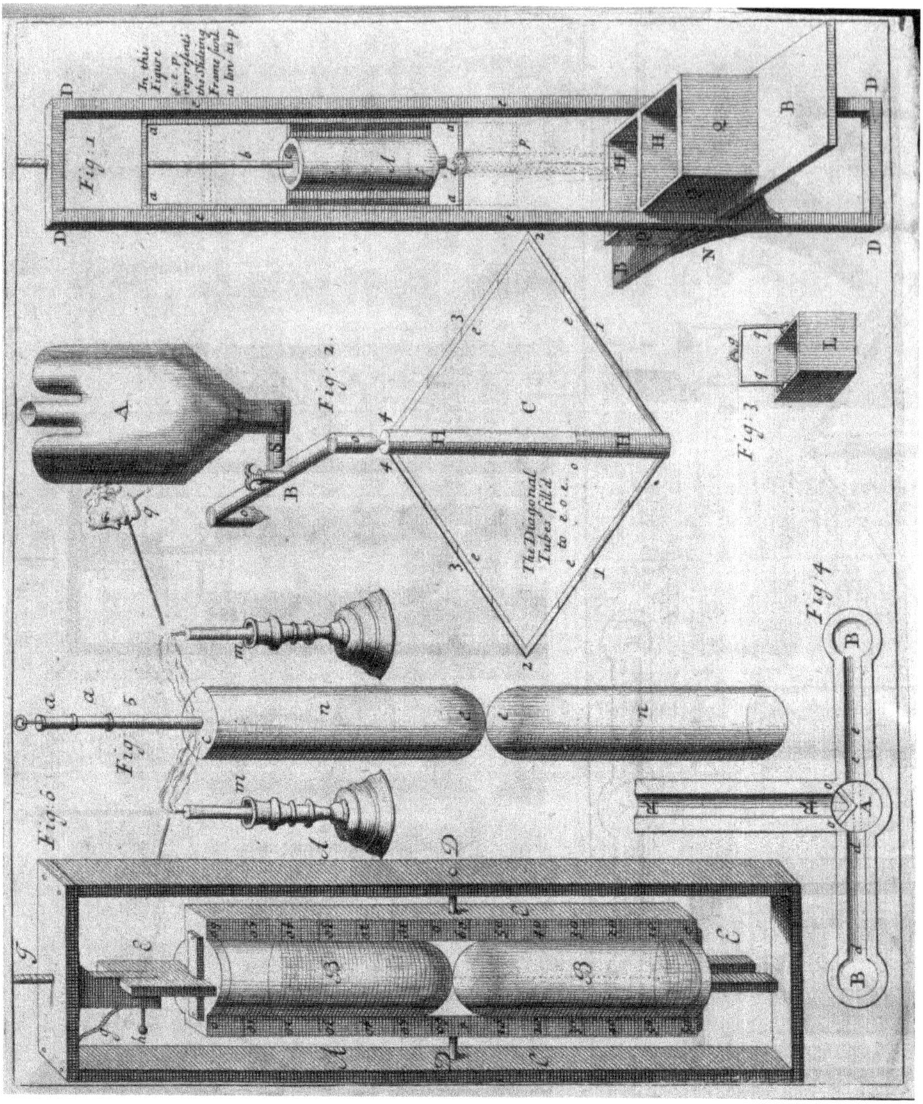

Figure 23: Plate from Clarke, *The mercurial chronometer improv'd* (London, 1715); engraving, 210 x 194mm, reproduced by kind permission of the Syndics of Cambridge University Library [P.19.25].

process leading to his invention, working across the white space of the page. Indeed, Thacker, Hobbs, and Clarke required their readers to relate images not only across a single page but across multiple pages and pamphlets. We see the illustrations in each of their exchanges working with the text to convince potential supporters of the relative success of their solution, and the comparative inadequacy of their rivals'. Thacker's choice of words to critique his rivals is also noteworthy, and we will return to the question of his identity in Chapter 3.

The Fluid Quadrant

One of the problems inherent in both Hobbs' and Clarke's illustrations is that of scale, in terms both of the relationship of different image sizes on the page, and of illustrating the measuring scale on the instrument itself. This was a visual problem that emerged in many longitude pamphlets, especially those that proposed observational instruments like quadrants. The problem was both in inscribing measurement scales onto the instruments themselves, and in making these visible when representing the instruments on paper. One of the first longitude proposals to present both sets of issues appeared in the 1720s, by the mariner Jacob Rowe (fl.1719–34), entitled *Navigation Improved: In Two Books*.[25] This was unusual in giving equal weight to latitude and longitude: each was the subject of one book, featuring a preface, explanation, and illustrations of how Rowe's instruments would work. The pamphlet's very form, therefore, recognised the importance of longitude and latitude working in relationship to each other, and like Whiston, Rowe emphasised the importance of accurate coastal charts.[26] The second half of the pamphlet dealt with the 'Everlasting Horometer' that Rowe proposed for finding longitude by using, like Clarke, the flow of mercury to measure time. This instrument had not been fully constructed and tested, and Rowe openly 'refer[red] to the Reader to examine and judge of the Theory, till proof is made, by publick Experiment'.[27] The illustrations here are numerous and deal with the different glass tubes that might be used to measure different periods of time (Figure 24, left side). Rowe, like Thacker, focused on these as the contested part of his invention, to which

25 This is unlikely to be the 'Mr Rowe' who proposed an instrument for observing lunar distance at sea to the Admiralty in 1703. His instrument was a nocturnal described by Flamsteed as 'one of the Worst Instruments that ever was Contrived for finding the hour of the night', and Rowe himself as 'aged and Necessitous', RGO 1/36: 17.

26 Rowe, *Navigation improved*, pp.ii–iv.

27 Rowe, *Navigation improved*, p.17.

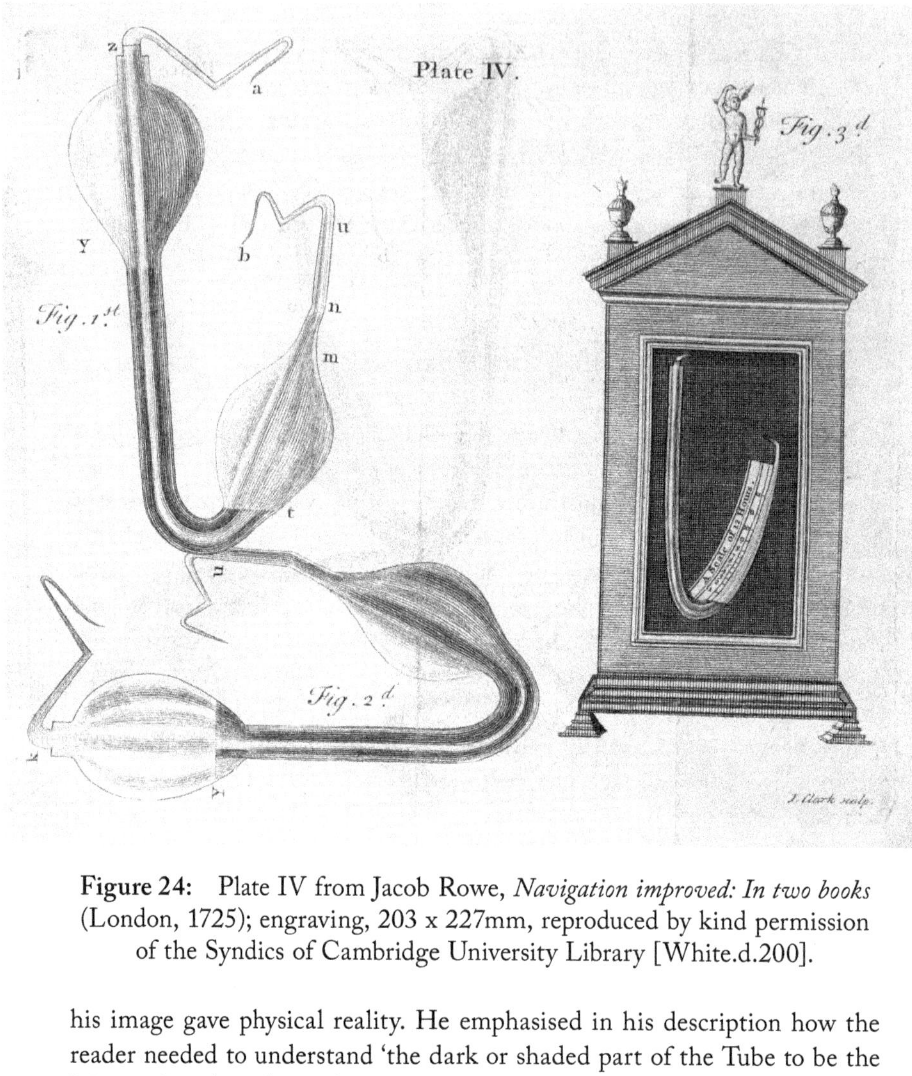

Figure 24: Plate IV from Jacob Rowe, *Navigation improved: In two books* (London, 1725); engraving, 203 x 227mm, reproduced by kind permission of the Syndics of Cambridge University Library [White.d.200].

his image gave physical reality. He emphasised in his description how the reader needed to understand 'the dark or shaded part of the Tube to be the Mercury', and used two figures one above the other to show different angles at which to hold the tube.[28] By contrast, the image of the whole instrument is much more generalised and figurative (Figure 24, right side), placing more emphasis on the decorative statue of Mercury than on the instrument illustrated inside the case. Here, we clearly see the problem of scaling. It is not clear what the size of the instrument, or indeed the case, is, and to give it physical reality the case has been extensively shaded and lit from the side. This creates a complex visual relationship with the instrument

28 Rowe, *Navigation improved*, pp.21, 27–31.

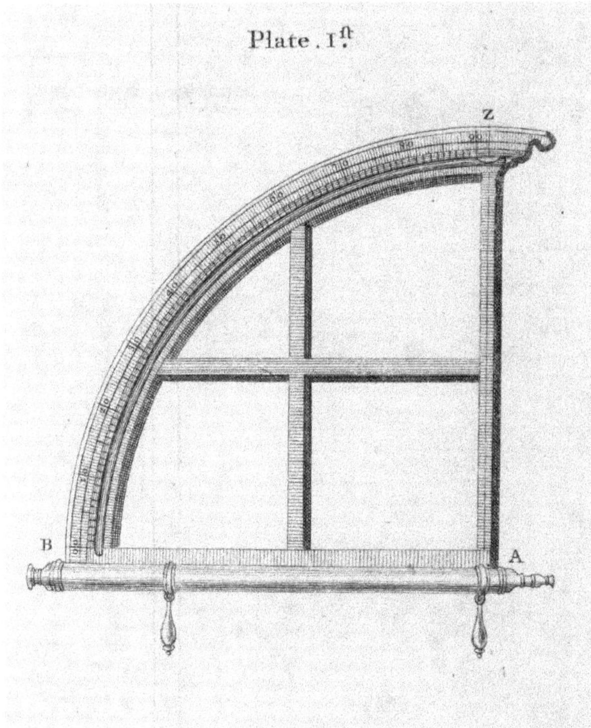

Figure 25: Plate I from Jacob Rowe, *Navigation improved: In two books* (London, 1725); engraving, 194 x 244mm, reproduced by kind permission of the Syndics of Cambridge University Library [White.d.200].

inside, which is barely shaded in order to be visible against the case, but on which the 'scale of hours' has had to be awkwardly curved to give an idea of three-dimensionality. While increasing the visual reality of the 'horometer', therefore, this markedly reduces its visual plausibility as a tool of accurate measurement.

This contrasts noticeably with the illustrations that accompany the first half of Rowe's pamphlet, which introduces the 'fluid quadrant' to measure latitude (Figures 25 and 26). This would make observation more accurate by using fluid in a curved tube as an artificial horizon against which to take observations. Rowe made explicit that he had constructed and tested this instrument, with reference to a carefully engraved illustration (Figure 25). Here his quadrant seems to come off the page, its shading and outline are so realistically achieved. Rowe explained how the engraved scale on the quadrant had been carefully tested to ensure accuracy, saying, 'the Degrees of the Arch [were] graduated by Experiment, by reason of some unavoidable Irregularities happening in the Glass Tube, which caus'd the Degrees in some Places to be uneven'.[29] He continued that he had subsequently developed the

29 Rowe, *Navigation improved*, p.xvi.

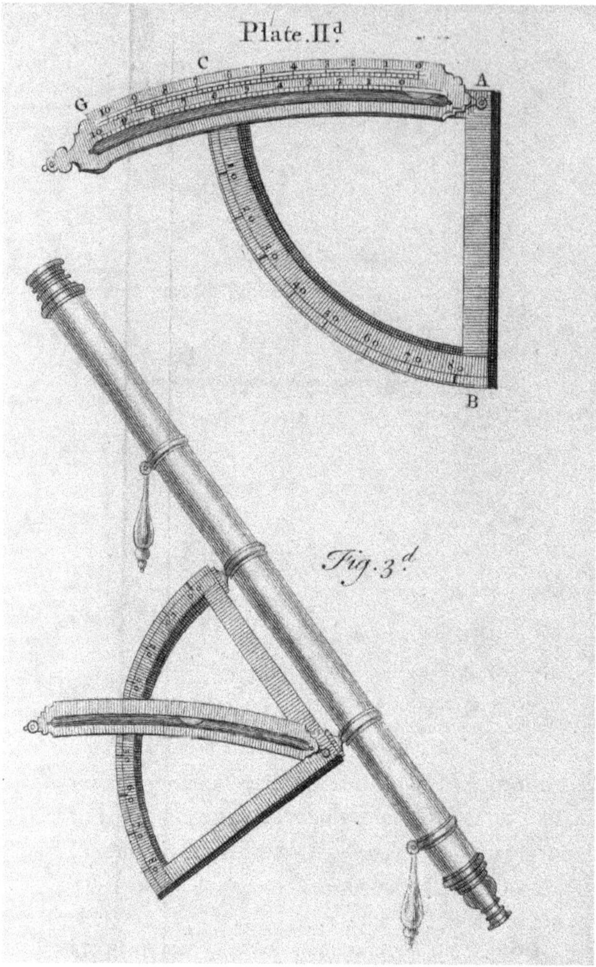

Figure 26: Plate II from Jacob Rowe, *Navigation improved: In two books* (London, 1725); engraving, 203 x 240mm, reproduced by kind permission of the Syndics of Cambridge University Library [White.d.200].

quadrant to make it more portable. In referencing the improved version in his second plate (Figure 26) he made a point of mentioning its size, advising, 'In Plate IId suppose the Quadrant A.B.C. six Inches more or less Radius',[30] and repeatedly explained how the movement of the fluid should relate to the 'Lines on the sliding Index' or on the Arch.[31] He used the two plates to demonstrate how the index arm should work, how to attach this smaller quadrant to a telescope, and how the removable handles should be attached. All three quadrant illustrations feature at least two carefully divided scales to show directly how observations would be measured accurately. Yet these

30 Rowe, *Navigation improved*, p.2.

31 Rowe, *Navigation improved*, p.7.

scales also create their own visual complexities. Each quadrant is shaded to give it physical reality, as we have seen, but the engraved lines that achieve the cross-hatched shading have the same spacing as the divides on the illustrated scales, making the latter, therefore, somewhat indistinguishable from the engraved contours. The illustrations add reality to Rowe's solutions, but the engraved grid that gives them substance also makes it harder to read the work done by the inscribed scale.

Rowe's contribution, too, was noticed and discussed by others. In the historical preface to the second edition of his work on magnetic dip, Whiston had commented on the accuracy of the fluid quadrant four years before Rowe published its description, referencing 'that newest and best Contrivance of the very Ingenious Mr. *Rowe*, whereby he can much more frequently, and somewhat more exactly than formerly take [the] Latitude'.[32] In the historical preface to his later 1738 proposal to use Jupiter's satellites, he commented that a clause should have been added to the 1714 Act for improving the means of finding latitude, outlining how,

> not very long after the passing of that Act, Mr Rowe procured a meeting of the Commissioners for the Longitude, in order to desire some Reward for an Improvement he pretended to have made in the Discovery of the Latitude ... altho' this Application appeared evidently beyond their Power, yet did he make his Pretensions so effectual at the Board of Admiralty, that a Bill was brought in.[33]

There had, indeed, been a bill brought to Parliament in 1719 on Rowe's behalf, which, although it did not pass through the Commons, shows the impression that his invention seems to have made on the Commissioners (although there is no record of the meeting described by Whiston).[34] It also suggests why Rowe later chose to publish and illustrate his invention, in an attempt to glean more widespread support. It had at least one demonstrable affect: on the Lancashire teacher and natural philosopher Robert Wright, who commented of Rowe's quadrant, in his *Humble Address* to the Commissioners in 1728, 'no Person that knows what is meant by the *Longitude*, can deny, but it may by this Means be found to the Exactness required by Act of Parliament, and a great deal less'.[35] He linked, in fact, both the definition of 'longitude' and the requirements of the Longitude Act to the success of the fluid quadrant.

---

32 Whiston, *Longitude and latitude found* (1721), p.98.

33 Whiston, *Longitude Discovered by the eclipses* (1741), p.vii.

34 *Journal of the House of Commons* (12–21 January 1719), pp.210–31.

35 Wright, *Humble Address*, p.19.

Rowe's naming of his instrument was also particularly apt. His quadrant was 'fluid' in the means he proposed to mark observations, but the quadrant was also more generally fluid in the range of developments proposed for it during the period. A plethora of pamphlets appeared critiquing and illustrating these, all aiming to convince readers of the reality and efficacy of the instrument there presented. Attempts to make a quadrant useable for accurate observations at sea had started with Newton, Halley, and Robert Hooke (1635–1703) in the late seventeenth century. Newton had at that point designed a reflecting instrument that he illustrated and sent to Halley, but which remained unpublished in Halley's papers until after his death. Halley used a version of this quadrant on the same Atlantic voyages on which he developed his magnetic variation maps.[36] The quadrant both entered the longitude debate and attracted public attention, however, in the 1730s with the work of John Hadley (1682–1744). His key improvement was the use of double reflection, which accommodated the movement of the ship at sea in taking sights, and therefore improved markedly on the backstaff.[37] Hadley published *A Description of a New Instrument for Taking the Latitude or Other Altitudes at Sea* in 1734, just as John Harrison was completing his first timekeeper ready for trial at sea. This pamphlet was a precise description of the instrument and how to use it, which Hadley aimed at instrument makers.[38] It included one simple but effective illustration, the key to which served to emphasise his central innovation that allowed the quadrant to take back as well as fore observations (Figure 27). He made sure that the reader understood the importance of the image, saying, 'the figure prefixed will better describe this Instrument than any Words alone can do'.[39]

Hadley went on to point out at whom this publication was aimed, adding, 'there are other Draughts of it in *Phil[osophical]. Trans[actions].* No. 420 with a Mathematical Account of the Principles it is founded on, and its several Properties and Uses, to which I chuse to refer the more Curious, this being designed rather for those not so much versed in Mathematicks'.[40] This directed the reader to the fuller, more detailed, and more comprehensively illustrated description of the quadrant that Hadley had presented to the Royal Society in 1731. The version of this that was published in the *Philosophical*

36 For the complex history of these early quadrants see Bennett, 'Catadioptrics and Commerce', pp.252–3, or Gerrard, *Astronomical Minds* for a more sensationalised version.

37 Bennett, 'Catadioptrics and Commerce', p.259. Newton's quadrant, unknown to all but Halley, had also used double reflection.

38 Bennett, 'Catadioptrics and Commerce', p.271.

39 Hadley, *Description*, p.6.

40 Hadley, *Description*, p.6.

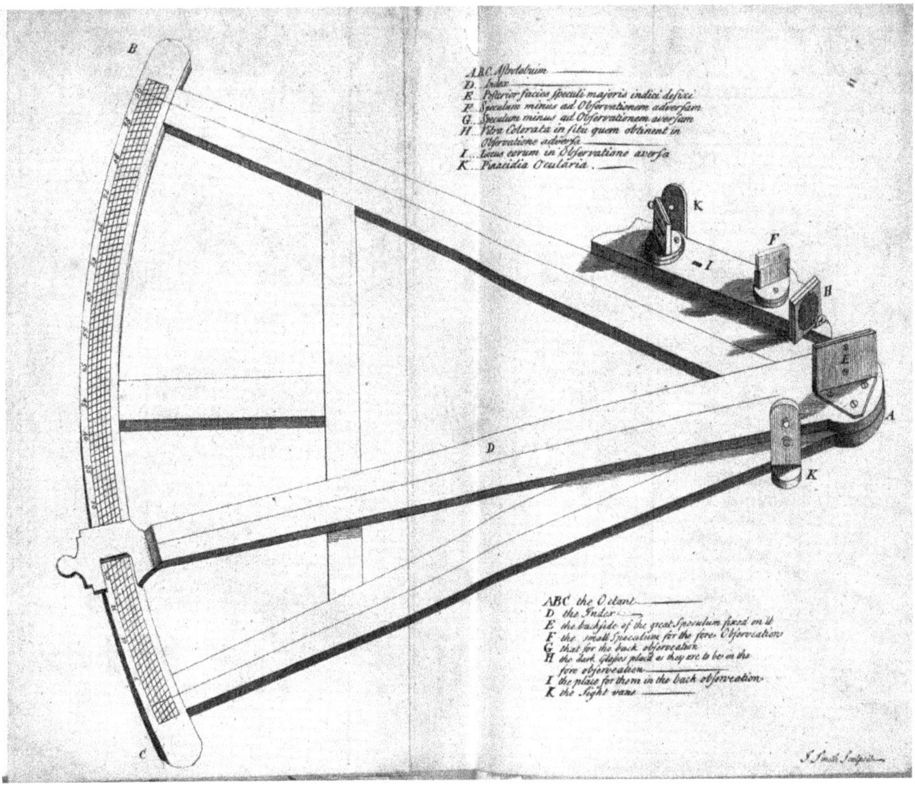

Figure 27: Quadrant illustration from John Hadley, *A description of a new instrument for taking the latitude or other altitudes at sea. With directions for its use* (London, 1734); engraving, 196 x 348mm, © The British Library Board (Shelfmark: 538.i.30.(7.)).

Transactions included two detailed illustrations, complete with diagrams of the geometry behind double reflection (Figures 28 and 29).[41] Like Rowe, Hadley included two sequential designs, the latter being the one published in 1734, which allowed for back observation and, therefore, angles larger than ninety degrees. Here, Hadley placed the telescope and glasses at a different angle, which allowed the frame to be held vertically, and therefore made it easier to use on a moving ship.[42] Both figures show detailed representation of the

41 Hadley, 'Description and Use of a New Instrument for Taking Angles' in *Philosophical Transactions* 37 (1731–32), pp.148–50. The figures reproduced here are from an abridged 1734 republication. The plates are slightly different but identical in the representation of the quadrant.

42 See Bennett, 'Catadioptrics and Commerce', p.260 and Morzer Bruyns, *Sextants at Greenwich*, p.25.

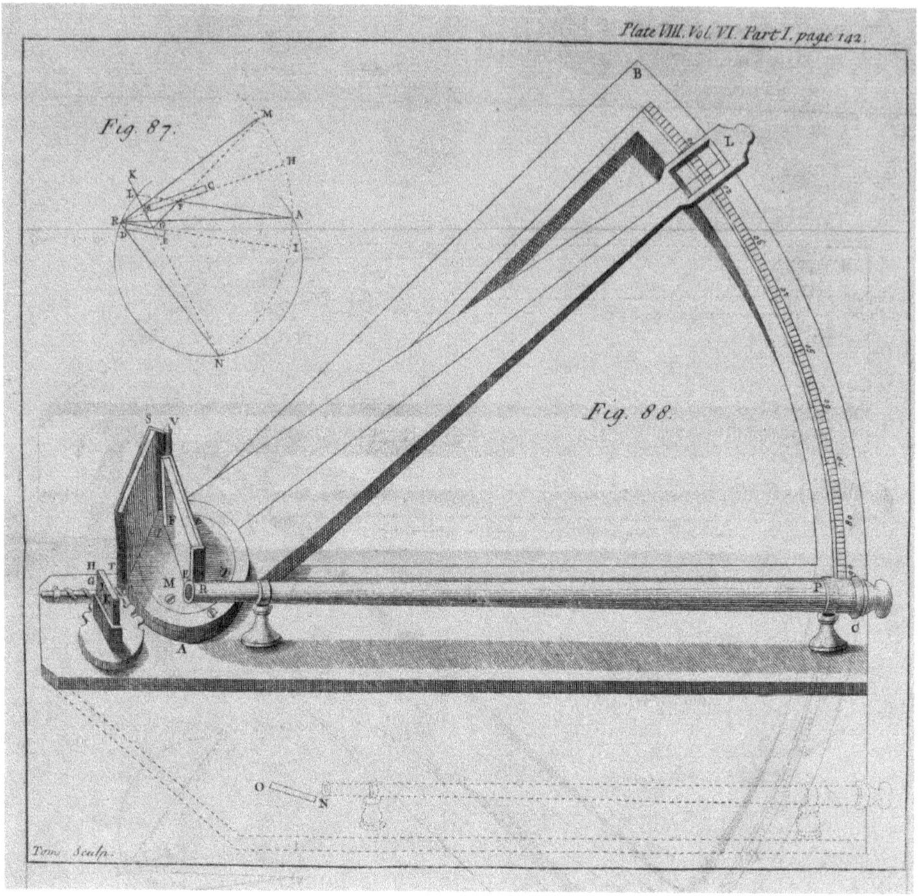

Figure 28: Plate VIII of 'The Description and Use of a New Instrument for taking Angles, by Reflection, invented by Mr. J. Hadley' in *The Philosophical Transactions (from the Year 1719, to 1733) Abridged, and Disposed under General Heads* (2 vols, London 1734); engraving, 220 x 197mm, Private Collection.

parts of the quadrant that created the double reflection, and have the index marked particularly carefully.[43] In a discussion of the innovations seen in the year 1732, *The Present State of the Republick of Letters* made clear to its readers the importance of looking at Hadley's images in the *Philosophical Transactions*, for 'it is impossible to give the reader a perfect idea of this instrument without a drawing, print, or some kind of representation upon paper'.[44]

43 Hadley, 'Description and Use', pp.150–1.

44 *Present State of the Republick of Letters*, ix, pp.239–40.

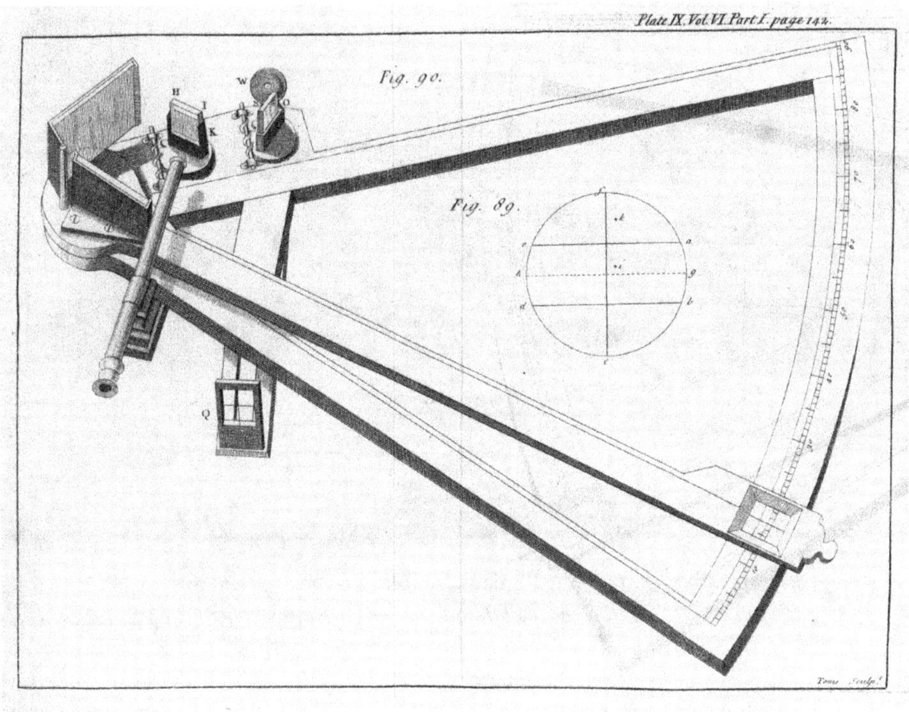

Figure 29: Plate IX of 'The Description and Use of a New Instrument for taking Angles, by Reflection, invented by Mr. J. Hadley' in *The Philosophical Transactions (from the Year 1719, to 1733) Abridged, and Disposed under General Heads* (2 vols, London, 1734); engraving, 193 x 260mm, Private Collection.

Later proposers of further developments to the quadrant clearly took this advice to heart. A steady series of pamphlets discussing improvements to the quadrant appeared from the 1730s to the 1770s, which frequently included figures of the instruments. Most immediately striking is a pamphlet published by the mathematician and instrument maker Benjamin Cole (1695–1766) in 1748, entitled *The Description and Use, of a New Quadrant, for Finding the Latitude at Sea*. In this, Cole discussed the history of the backstaff, otherwise known as 'Davis quadrant', after its inventor John Davis (c.1550–1605). He emphasised how his own improvement to the invention 'arises from the manner of its divisions; whereby the accuracy of the degrees, and parts of degrees on the quadrantal arch, may easily be examined'.[45] He went on to discuss Hadley's quadrant in detail, explaining its different parts

45 Cole, *Description*, p.19.

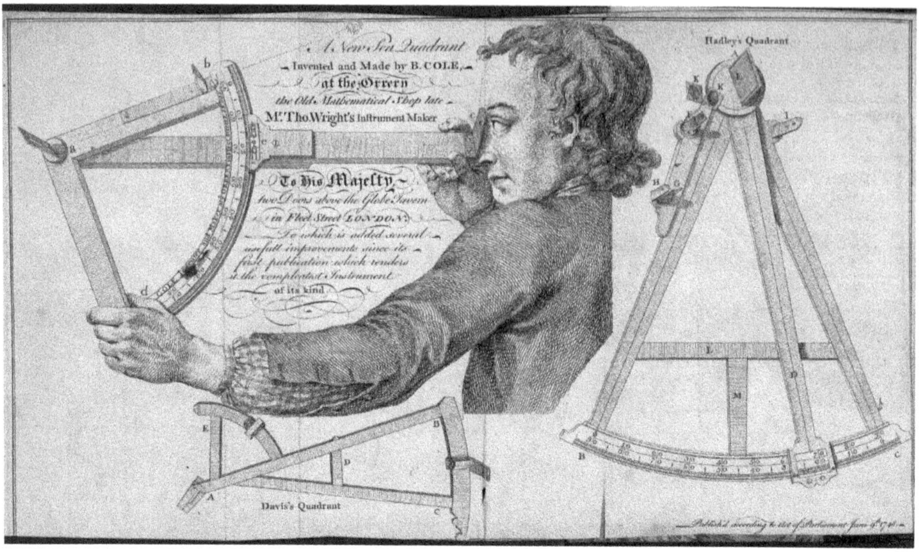

Figure 30: Plate from Benjamin Cole, *The description and use, of a new quadrant, for finding the latitude at sea* (London, 1748); engraving, 197 x 245mm, © The British Library Board (Shelfmark: 1651/524).

and how to use it at sea.[46] What is truly remarkable about this pamphlet is the image (Figure 30), which very unusually shows a man in the act of using Cole's quadrant, not in a frontispiece, but in the technical illustration of the instrument itself, reminiscent of the hands and heads seen at work in Clarke's second pamphlet of 1715. This mariner is shown wedged into the centre of the plate, with his back to the image of Hadley's quadrant, and his arm curving above the image of the backstaff. His eye is to the sight vane of his quadrant, with his strongly moulded hand drawing the viewer's eye to the clearly marked scale on the index. Above the index, Cole's name and credentials are inscribed into the scale. Cole's image, therefore, not only works to illustrate his discussion of the use of the various quadrants, but also creates a visual history of their use and development. The viewer can follow the innovations by Davis, Hadley, and Cole anticlockwise around the figure. We will consider the wider significance of such figures in instrument illustrations in Chapter 5.

In 1772, the famous London instrument makers father and son John (1706–61) and Peter Dollond (1731–1820) took these ideas a step further in their *Directions for Using the New Improved Hadley's Quadrant*.[47] They

46 Cole, *Description*, pp.23–32.

47 Peter and John traded as P. and J. Dollond from 1766.

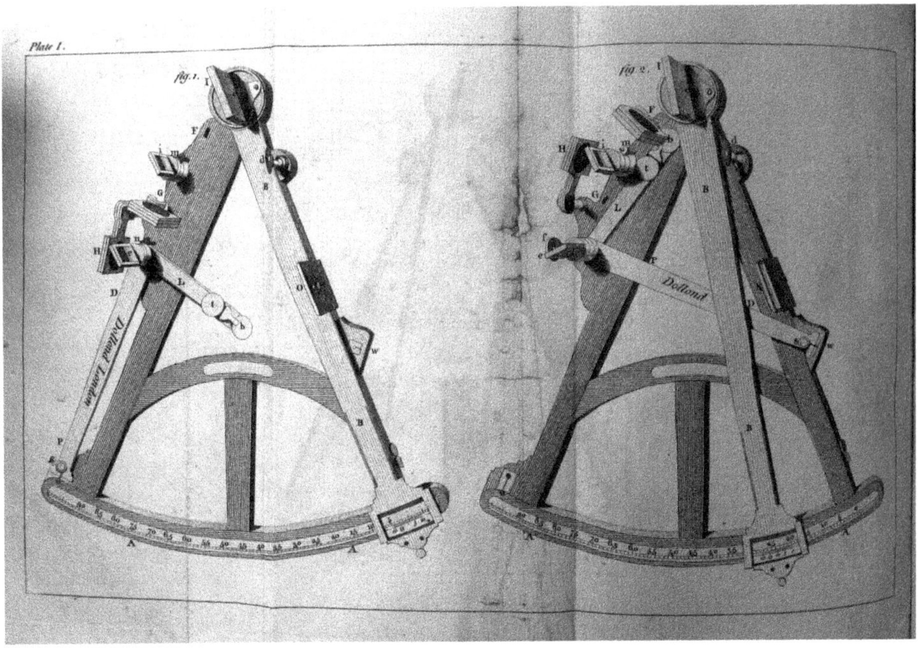

Figure 31: Plate I from P. and J. Dollond, *Directions for using the new improved Hadley's quadrant; with the new method of adjusting the glasses for the back observation* (London, c.1772); engraving, 210 x 290mm, by permission of the Master and Fellows of St John's College, Cambridge.

emphasised how their improvements to the glasses allowed back observations to be made more accurately and easily. They used their illustrations as a means to demonstrate how to adjust the quadrant between fore and back observations, making direct comparisons between different figures on the same plate (Figure 31). Thus they explained, for instance, how to adjust the glasses, saying, 'when these glasses are used for the fore observation, they are fixed at *F*, fig. 2, when used for the back observation, they are removed to *G*, as in fig. 1',[48] or how to move the index to read off these observations: '*P D*, fig. 1 and 2, is an index that turns on the centre, over which the back horizon glass is fixed. By moving this index from the position shewn in fig. 2, to the position shewn in fig. 1, the back horizon glass is brought into a position like to that of the fore horizon glass'.[49] The scale is carefully described on both quadrants. The Dollonds also separated out the significant parts of their quadrant in a second plate (Figure 32). With reference to this,

48 Dollond, *Directions*, p.9.

49 Dollond, *Directions*, p.19.

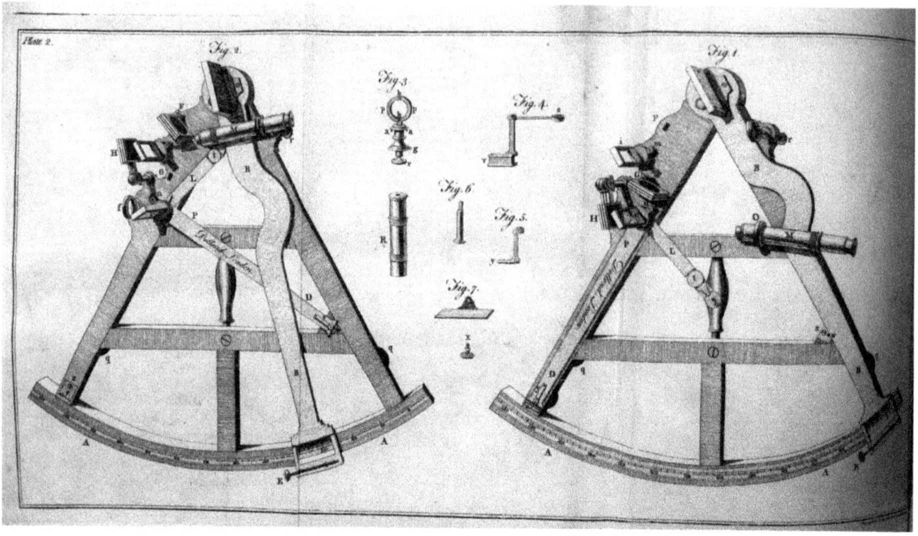

Figure 32: Plate II from P. and J. Dollond, *Directions for using the new improved Hadley's quadrant; with the new method of adjusting the glasses for the back observation* (London, c.1772); engraving, 214 x 340mm, by permission of the Master and Fellows of St John's College, Cambridge.

they discussed how to adjust the telescope for different observations, saying, 'when it is used for the fore observation, it is screwed into the circular ring as is shewn fig.2; and when used for the back observation … screwed into the other circular ring at *f.* The part that supports the telescope when used for the back observation is represented at fig. 3'.[50] Thus, they not only used their two images to demonstrate use of the instrument but referenced the illustrated small parts into this process. We notice that the signature 'Dollond London' is prominently inscribed on the index arm of every quadrant (Figure 31). The Dollonds, like Cole thirty years earlier, made sure the image of their invention could not be separated from its association with their name.

'By his Majesty's Royal Patent'

In inscribing their signature into the adjustment lever of their quadrant, the Dollonds made a forceful statement for their ownership of its design, as well as of the illustrated plate. The title page of their pamphlet also made their ownership claim clear in stating that their method of adjusting

50 Dollond, *Directions*, pp.35–6.

the glasses was 'by his majesty's royal patent'. They affirmed a legal right to the sole ownership of this invention. Patents had grown out of the concept of royal monopolies, and had, therefore, acted as the model for copyrights in both the authored texts and engraved images discussed in the previous chapter.[51] The concept of patents and the steps required to be awarded one changed markedly over our period, in line with changes in wider copyright legislation.[52] They were also intimately tied to the longitude problem as they frequently focused on the two types of invention we have been looking at as central to the longitude debate: clocks and navigational instruments.[53] It was arguably interest in the longitude problem, and particularly the 'great reward', that was responsible for encouraging patenting and expanding its application to other instruments.[54] Many of our longitude proposers applied for patents, including Rowe, Hadley, Christopher Irwin (fl.1762–65)—inventor of the marine chair—and the Dollonds.[55] Peter Dollond's insistence on applying his father's patent for achromatic lenses in 1758 is generally seen as instrumental in bringing the law into practice. As we have seen with their 1772 images, the Dollonds used the patent as a tool for trademark and branding, and then to protect their intellectual copyright against other makers.[56] Some patents were used as a means of quality control; Hadley used his, for instance, as a means of establishing his priority, but then of licensing other makers to ensure the accuracy of his quadrant.[57]

Yet this is not just a story about instruments. We can see a spectrum from print to instruments in this chapter, in the development of legislation around copyrights, in ideas about engraving scales as well as prints, and in concerns about how to represent cartographic ideas or newly invented instruments on paper. Hogarth succeeded in getting the Engravers' Act passed in 1734–35 as a product of his concerns about his own rights over his paper products. But this was also precisely the moment at which proper specifications started to be mandatory for patent applications, although what that specification should

51 Rose, 'Author as Proprietor', p.61.

52 Macleod, *Inventing the Industrial Revolution*.

53 Biagioli, 'From Print to Patents', p.146. David Bryden argues, however, that, whereas many instrument patents are concerned with navigation, they are not predominantly concerned with longitude (Private Communication).

54 Biagioli, 'From Print to Patents', p.167.

55 Irwin proposed a chair that would allow the sitter to take steady observations of Jupiter's satellites on shipboard. See Irwin, *Summary*. This was trialled by Maskelyne en route to Barbados, where he also participated in the second trial of H4. He concluded it was useless. See Meeting of the Board of Longitude, 18 September 1764, RGO 14/5: 67–8.

56 Biagioli, 'From Print to Patents', pp.143–4.

57 Bennett, 'Catadioptrics and Commerce', p.269.

include would require further legislation to determine.[58] Patents developed precisely alongside new ideas about literary copyright, about visual copyright, and about the importance of the media of representation in processes of disclosure. Patenting became viable only 'when patent lawyers and administrators had access to the specific material media in which inventions could be represented; namely, scale models, formalised texts, technical drawings, and archives'.[59] Patents were established through lines on paper. They were carefully defined through words at the same time as dictionaries and encyclopaedias developed, and they required inscribed specifications at exactly the moment that Hogarth acquired copyright for engraved images. From the 1730s, specifications shaped what could be patented, such that inventions needed to be inscribable in words and images on paper. In accusing illegal copyists of 'piracy', as we will see Harrison did, intellectual property also became specifically maritime in terminology.

JOHN HARRISON'S 'CORRECT DRAWINGS AND DESCRIPTIONS IN WRITING'

BY THE 1770S, the drawings involved in patent processes were starting to become highly skilled as the primary method of communicating inventions.[60] John Harrison also became intimately involved in the complex project of delineating paper schemes such that they could establish priority and property. His interaction with the Board of Longitude, in the 1760s and '70s, concerning how to publish the workings of his watch helped to develop the ideas behind both the patenting and copyright systems. We have seen how they argued over the definitions of the 1714 Act, especially in the process of passing a further Act in 1765. This was not only over the outcome of the trials, it was also over how the success of Harrison's watch could be made both transferable as an inscription upon paper, and mobilizable as an instrument that could be made available to the Navy.[61] In the process of disclosure, Harrison was constantly concerned to ensure his continued rights, while the Commissioners essentially worried over the accuracy of the 'specification' that he had given them. In 1778 in the case of *Liardet v. Johnson*, Lord Justice Mansfield (1705–93) would decide that a specification had to provide

58 Pottage and Sherman, *Figures of Invention*, p.52.
59 Pottage and Sherman, *Figures of Invention*, p.2.
60 Macleod, *Inventing the Industrial Revolution*, p.49.
61 This idea of mobility builds directly on Latour's concept of 'immutable mobiles' in 'Visualisation and Cognition'.

sufficient detail for a member of the relevant trade to use and apply it without further explanation. He made the disclosure essential to gaining a patent.[62] This is the same process that Harrison had only recently negotiated in order to gain his equivalent of a patent: the longitude reward.[63]

In his concerns over the safety of his ideas and his rights over them, Harrison showed the same worries about quality control, priority, and branding that we have seen utilised by Hadley and Dollond in patenting their quadrants. It is significant how he repeatedly returned to ideas of intellectual 'piracy' in concerns about ownership of his ideas.[64] In the *Account of the Proceedings, in Order to the Discovery of the Longitude* submitted to Parliament to petition for the unsuccessful 1763 Act, Harrison discussed the history of his inventions. He told how, in 1749, when he was awarded the Copley Medal by the Royal Society, President Martin Folkes (1690–1754) had given a speech at the presentation ceremony

> shewing the Construction of the Instrument, and various and curious Inventions used to bring the same to Perfection: Which ... gave an Opportunity to Pirate such Parts of the Invention, which the low Mechanicks were capable of Understanding, particularly the Gridiron-Pendulum. / And which ungenerous Treatment, made it necessary for Mr. *Harrison*, to be cautious in shewing his Inventions.[65]

In the build-up to this Act, Harrison wrote to the Board making his growing caution clear and proposing his terms for giving a disclosure, 'provided, that before the said invention shall be ... made publick the property of the same shall be secured by Parliament to me'.[66] Indeed, in his *Remarks on a Pamphlet Lately Published by the Rev. Mr Maskelyne*, Harrison practically accused the Commissioners themselves of pirating his work from the disclosure, complaining, 'they have since published all my Drawings without ... even paying me and my Son for our Time at the Rate of common Mechanicks'.[67] It is the complex process of bringing these drawings to publication to which we now turn.

62 Macleod, *Inventing the Industrial Revolution*, p.49.

63 William Harrison himself later sought a patent for a method of making screws, without a drawing, just before Mansfield's decision. He applied with Peter Atherton on 5 February 1778; see Woodcroft, *Alphabetical Index*, p.1, 179. The *Morning Post* reported Mansfield's decision on 23 February 1778, Issue 1,667.

64 Johns, *Piracy*. Quill also commented on Harrison's use of the word in *The Man who Found Longitude*, p.146.

65 Harrison, *Account*, p.22.

66 *HJ*, p.36.

67 Harrison, *Remarks*, p.22.

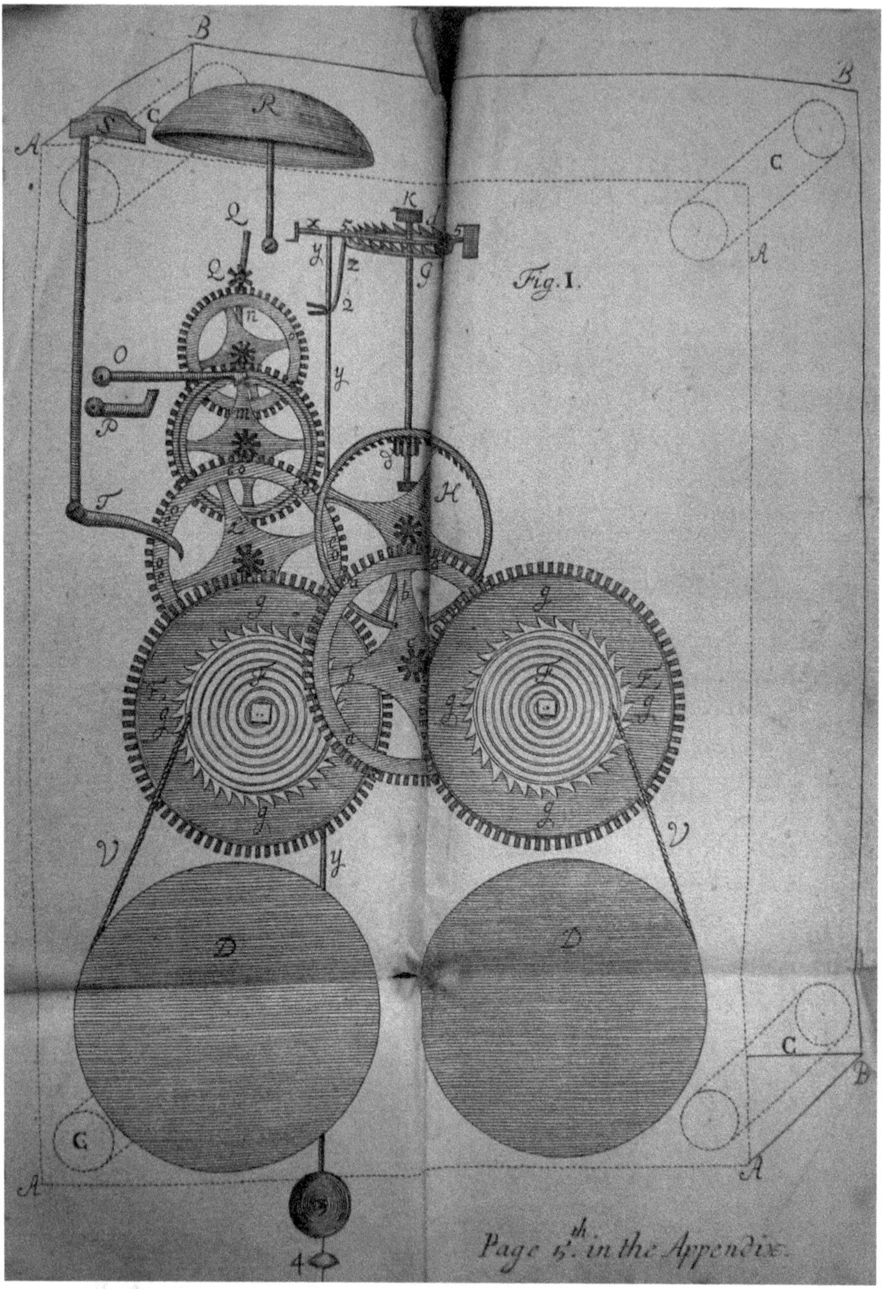

Figure 33: Plate from William Derham, *The Artificial Clock-Maker* (London, 1714); engraving, 196 x 152mm, Whipple Library, University of Cambridge.

The different processes of illustration and inscription that we have been discussing, then, are perfectly effective when you need to see only the outside of an instrument. Quadrants could be easily illustrated to show the key parts of their operation, as these were external. Timekeepers were not so easy. We have seen how Hobbs, Clarke, and Thacker all struggled to create a convincing image of the workings of their chronometers, and especially the tension in Hobbs' images between a decorative clock face and poorly visualised mechanism (Figures 18 and 19). We might compare this to another work from 1714, by William Derham (1657–1735), theologian, Fellow of the Royal Society, and horological enthusiast. He published *The Artificial Clock-Maker* soon after the passage of the Act, the third edition of a work also published in 1696 and 1700.[68] This was a practical treatise explaining how to perform the calculations necessary to construct a timekeeper, and how the different parts worked, with special emphasis on the pendulum. It is noteworthy that the 1714 edition was both the first version of the text to use the term 'chronometer', and the first to discuss how a timekeeper could be used to find longitude at sea. The sole figure that Derham included in both the 1700 and 1714 editions (Figure 33) showed a carefully shaded illustration of the inner workings of the clock: the wheels, pinions, chimes, and pendulums that he discussed in the text. These parts are superficially contained within a case marked out with dotted lines, but these are as insubstantial as Mountaine and Dodson's dotted magnetic variation lines. Against them, the clock's innards leap out from the page, convincingly solid and functional. This gives them reality, but it does not create the diagrammatic type of illustrations needed for others to copy, in specifications, or by the Board's definition of the Longitude Act.

Harrison was clearly accustomed to processes of drawing and used these as part of his work in developing the timekeepers. An account of his life published in a number of contemporary newspapers in 1778 told the story of how he had first become interested in mechanics, 'frequently employing all, or great part of his nights, in writing, or drawing … he always acknowledged his obligations to a clergyman … who lent him a[n] MS. copy of Saunderson's Lectures; which he carefully and neatly transcribed, with all the diagrams'.[69] This was most likely also the source from which Harrison first learnt of the Longitude Act itself, so that the development of his ideas came out of processes of drawing at the same time as he first encountered the longitude problem.[70] Two rough drawings survive from the 1730s when Harrison

68 Derham, *Artificial Clock-Maker*.

69 'Account of the life of the late Mr. Harrison' in the *Annual Register* (1778), p.25.

70 Hogarth also mentioned the blind Saunderson in the *Analysis*, as exemplary of the power of feeling in aiding visual experience, p.84.

was developing his first marine timekeeper, in which he described his work on clocks thus far and the level of success of his various innovations.[71] He used the figures as a means of illustrating the text, including key parts of the instruments such as his gridiron pendulum and pair of balance springs, but also discussing key experimental points in his work with, for example, different metal wires. He therefore presented his inventions in just the same way as Thacker, and then Clarke, had sixteen years earlier, with a mixture of diagram and illustration to give visual plausibility. In presenting the first proposal to be seriously considered by the Board of Longitude, Harrison brought such questions of illustration into focus. The success of his timekeepers forced the Commissioners to consider the ways in which they needed proposals to be recorded and communicated on paper in order for Harrison to qualify for the 'great reward'. Their problem with Harrison was one aspect of those discussed so far, shared with the adjudicators of patent applications, or critics of longitude pamphlets. In forging the relationships between Harrison's watch and the paper records that they required, the Board of Longitude created new mechanical drawing for clockwork. The diagrams took the viewer inside the watch-work in an unprecedented way, in an attempt to make both Harrison's instrument and his mental framework behind it legible both to the Board and, more importantly, to other clockmakers.

The debates that surrounded the publication of *The Principles of Mr Harrison's Time-Keeper* in 1767 were lengthy and complicated. The Commissioners were very clear about what information they wanted, and how this needed to be recorded on paper. They requested,

> 1st, That in case there are any particular Tools necessary towards the construction of Mr Harrison's Watch ... that Specimens be exhibited & correct drawings and descriptions in writing be given of such Tools ...
>
> 2ndly, That correct drawings and descriptions in writing be given of the several parts of Mr Harrisons Watch with a clear and full explanation of their power and of the manner in which they act with respect to one another.[72]

They clearly required the key information relating to the workings of the watch, and to the process of making it, both to be written down and drawn. They wanted both a textual and visual product for publication, and they wanted to understand clearly how the different parts referenced each other. Harrison, by contrast, placed much greater emphasis on oral communication. He suggested that,

71 'Description, with plans', Harrison Papers, CLC/L/CD/E/042/MS06026/001.
72 *HJ*, pp.40–1.

> 1st I will shew [*sic*] and explain what Tools are necessary for making such [a] Timekeeper in so distinct a manner that other Workmen may be able to make the same and this I will do by talking; for drawings are not necessary for such purposes.
>
> 2[ndly] I will shew [*sic*] drawings of the Timekeeper and of all its parts and explain the manner in which they Act with respect to one another this I can do in a much fuller and clearer manner by speaking than by writing.[73]

He showed an ambivalent attitude to drawings as a means of communication. He specifically questioned the usefulness of drawings for recording his use of tools, and he offered to show drawings but by no means to 'give' them as the Board requested. He also questioned the link between such drawings and written descriptions contending that these would be communicated more clearly verbally. Both Harrison and the Commissioners appreciated that the communication needed to be made in such a manner that other workmen could make the watch, but the Commissioners required this in a drawn form that they could publish, while Harrison saw the need for a practical communication to other makers. We have seen that he feared for his control over his invention once this was written down and transportable, but he seems also to have doubted, like the Commissioners, that his knowledge could be recorded in a visual way.

The publication itself shows that the Commissioners' concerns over accurate drawing continued. The *Principles* is introduced by a 'Preface' and 'Notes' from the Commissioner and Astronomer Royal Nevil Maskelyne (1732–1811), who attended the disclosure, in which he added his own discussion of the description and drawings. He added more detailed information about the workings of certain parts of the watch and anecdotes on its development and trials, giving temporal substance to the watch's claim to accuracy. He also added comments from other instrument makers, John Bird (1709–76) and Thomas Mudge (1715–94), as to certain mechanical operations.[74] Most importantly, he commented extensively on the annexed plates and how they had been engraved from the drawings, saying,

> The original drawings of the watch were delivered to me at the board of longitude, which I have since caused to be carefully engraved, under my own eyes ... and think I can answer, that the lines on the copper-plates every-where correspond with the measure of the drawings within one, or at most two breadths of the lines, which are so very fine.[75]

73 *HJ*, p.44.
74 *Principles*, pp.xiv–xv.
75 *Principles*, p.vi.

We note the emphasis both on lines and on the accuracy of engraving. Maskelyne made clear the margin of accuracy of the engravings, in terms both of the copying from drawn to engraved formats, and the accuracy of the engraving process itself. Translations of the *Principles* in French and Danish appeared with astonishing speed in the same year, where both Esprit Pezenas (1692–1776) and Christian Carl Lous (1724–1804) likewise made a point of discussing how their engravings related to the originals.[76]

This was also the period in which debates were carried out over the quality of British paper manufacture.[77] Maskelyne went on in his 'Preface' to comment specifically on the accuracy of the impressions of the engraved plates on the paper, saying,

> the reader must not expect to find the impressions taken off upon common paper agree so nearly, either with the originals or with themselves, owing to the shrinking of the paper after the water has been pressed out of it in going through the press. But, for the sake of the curious, and particularly artists who may be desirous to construct other watches after the model of Mr Harrison's, I have caused a few impressions to be taken off upon India paper.[78]

He considered the physical process of transferring the lines to the paper and made it very clear that the Board were seeking to publish drawings that could be used by 'artists who may be desirous to construct other watches'. Thus, in this sense, the engraved drawings became specifications of the watch.

There were fifteen engraved figures on ten plates accompanying the *Principles*, nine of which are the same size as the pages of text, and one of which folds out to double size (illustrated here are Plates I, II, IV, IX, and X as Figures 34–38). The figures involve no shading or perspective construction, no image of the instrument as a whole, and no image of the dial plate, despite the fact that different designs for the dial by Harrison exist with the original drawings.[79] Some plates include both front and side views with dotted lines showing where the views interrelate (e.g., Figure 36). They are simple, diagrammatic line drawings, with extensive key lettering that links to the 'Description'. Besides this lettering, the only writing on the plates discusses the size of the different parts, or slight inaccuracies in

76 *Principes de la montre*, p.22. See Turner, 'Berthoud in England', pp.219–38. *Historien af Mr. Harrisons Forsøg* (with thanks to Ivan Tafteberg for the translation).

77 Fox, *Arts of Industry*, p.198.

78 *Principles*, pp.vi–vii. Financial accounts in the Board archives show that this use of different papers to ensure accurate printing of Harrison's plates was still carried out as late as 1819 (RGO 14/19).

79 'Mechanical Notes', Harrison Papers, CLC/L/CD/E/038/MS03972/003, pp.57–9.

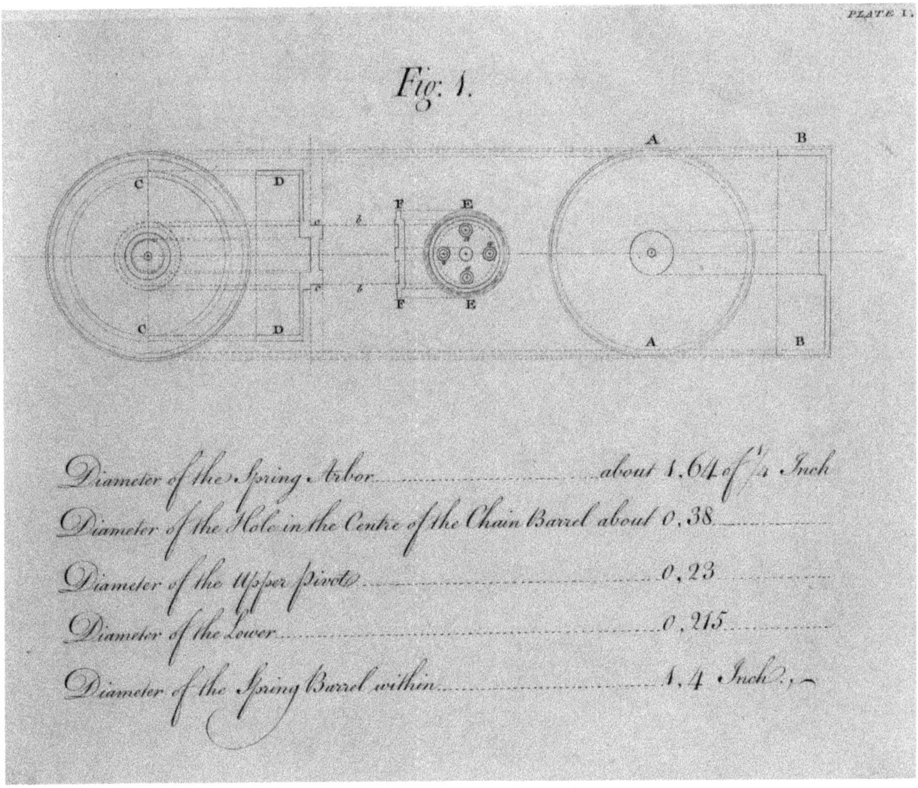

Figure 34: Plate I of *The Principles of Mr. Harrison's Time-Keeper* (London, 1767); engraving, 240 x 280mm, Royal Museums Greenwich, London.

the images (Figure 36). These notes on sizing add to the drawings' role as to-scale copies of the watch from which other makers could cut their parts, transferring the inscription on the page directly to the mechanism. The 'Description' certainly emphasises that these are drawings 'from which' the watch was made, and it seems clear that Harrison was unusual in the extent to which he worked out his ideas on paper before putting them into practice.[80] Jonathan Betts (b.1955) argues, nonetheless, that these plates were sufficient to enable other makers to copy the watch. Some parts are not entirely clear, but the key element that keeps accurate time, the oscillator, is carefully drawn and explained and had an observable impact on clockmakers.[81] It was not only the scale of the parts, but the scale and exactness of these engraved

80 *Principles*, p.22.

81 Personal communication from Jonathan Betts, and his 'Arnold and Earnshaw' in Andrewes, *Quest*, pp.311–28.

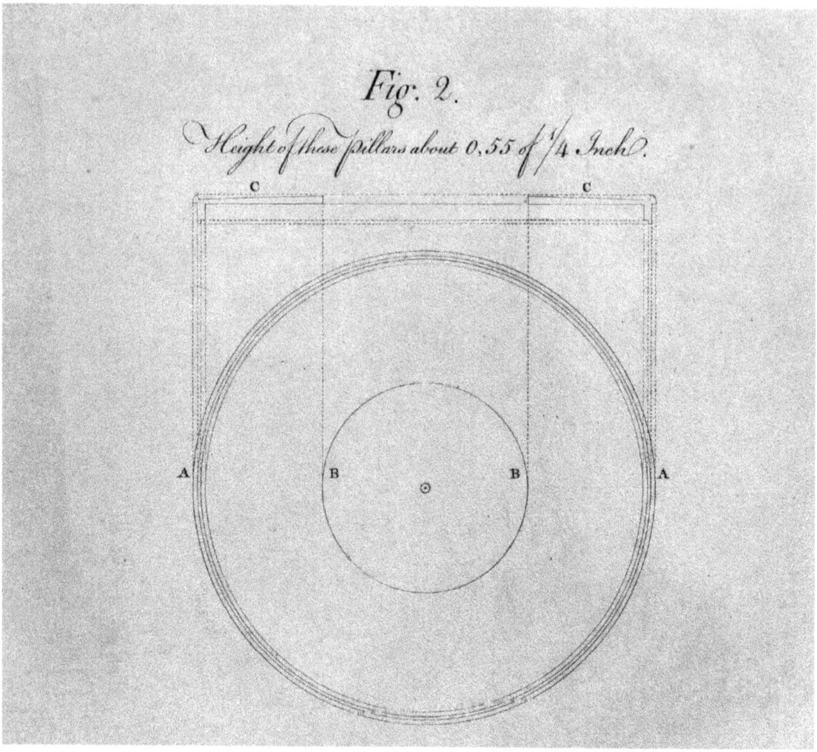

Figure 35: Plate II of *The Principles of Mr. Harrison's Time-Keeper* (London, 1767); engraving, 240 x 280mm, Royal Museums Greenwich, London.

drawings that allowed this impact. These were unique in horological literature at the time, as we can see by comparison with Derham's plate (Figure 33), one of the nearest comparable 'practical' works.

The 1767 engravings are, in fact, the perfect example of the work done by engraved lines on paper for which this and the previous chapter have argued. The plates and figures work in layers, referencing each other into the accompanying description. Plates I to VIII build up images of the crucial small parts (e.g., Figures 34–36), which the descriptive text relates into the constituent images of layers of the watch in Plates IX and X (Figures 37 and 38). These two last plates show figures layered in themselves, featuring multiple overlapping parts, which the earlier plates illustrate and explain independently. The text makes references between these with the complex lettering key. For example, in describing Plate II (Figure 35) the 'Description' explains, 'AA is the Brass Edge ... This Brass Edge is supported by six Pillars, and their places are represented in Figures 13 and 14, by six Circles *aaaaaa*'

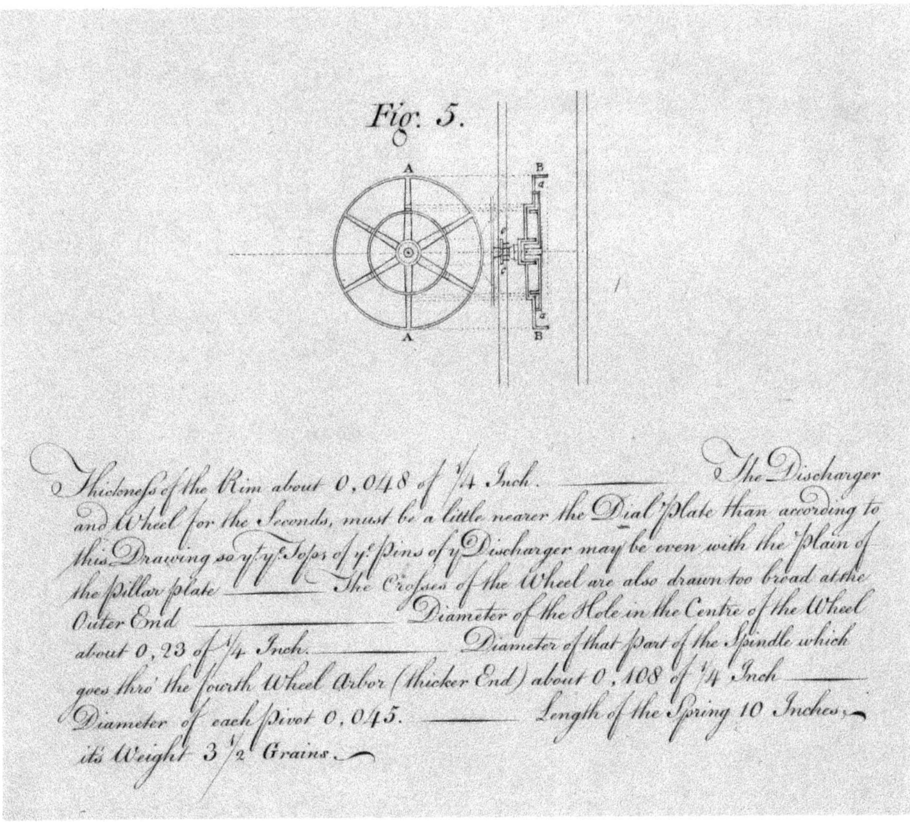

Figure 36: Plate IV of *The Principles of Mr. Harrison's Time-Keeper* (London, 1767); engraving, 240 x 280mm, Royal Museums Greenwich, London.

(Figures 37 and 38).[82] The drawings and 'Description' create a complex network of references between the different figures, and between the figures and the text. Maskelyne's 'Notes' then reference these securely to the original object through his own observations. Yet the process of reference within the text is problematic. To understand the complex relationship between the different parts of the watch and the text, the reader in fact needs to approach the *Principles* backwards. The text is presented with Maskelyne's 'Notes' and 'Preface' as the gloss on Harrison's 'Description', which is, in turn, the gloss on the plates. But none of these glosses has a point of reference unless the image is read first. Thus, the reader really needs to look at the plates, starting from the final composite ones, and working backwards to the specific parts,

82 *Principles*, p.22.

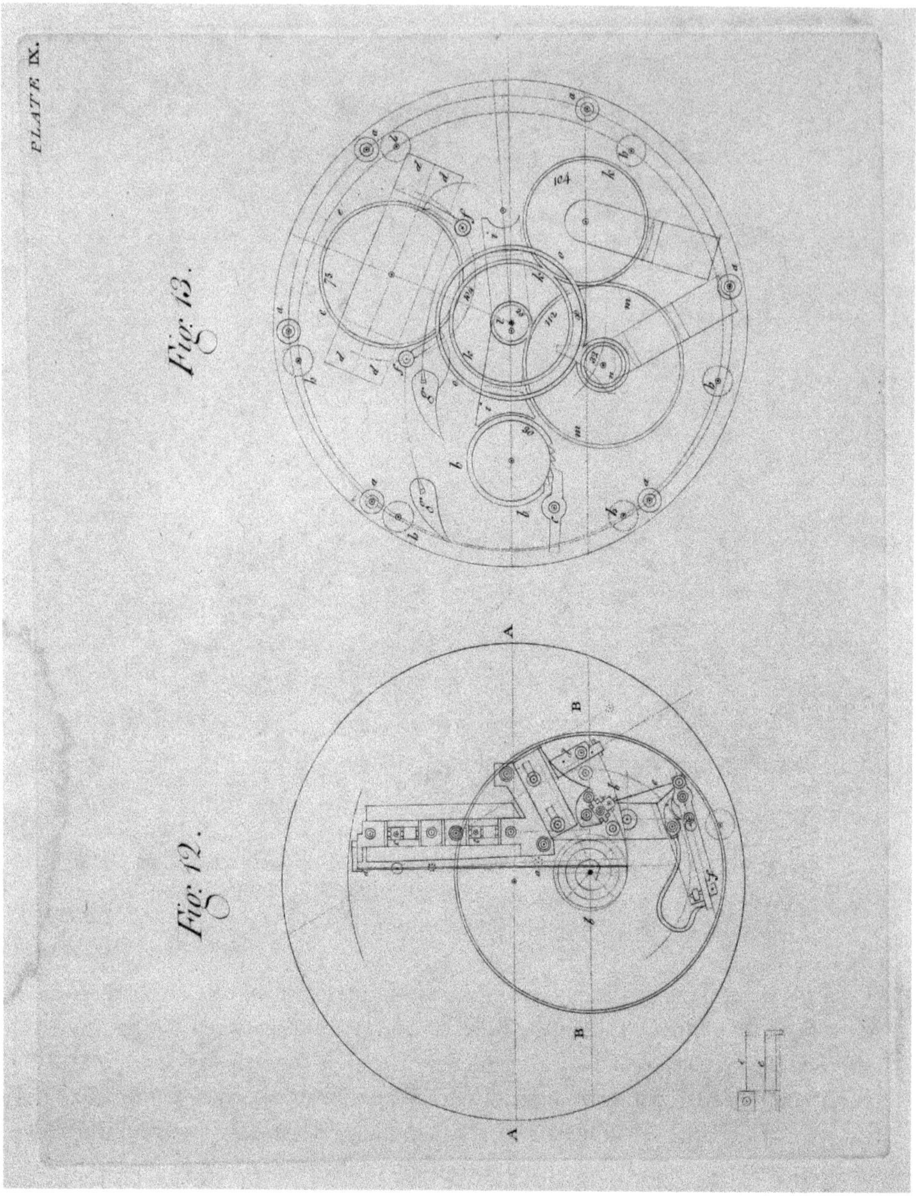

Figure 37: Plate IX of *The Principles of Mr. Harrison's Time-Keeper* (London, 1767); engraving, 240 x 280mm, Royal Museums Greenwich, London.

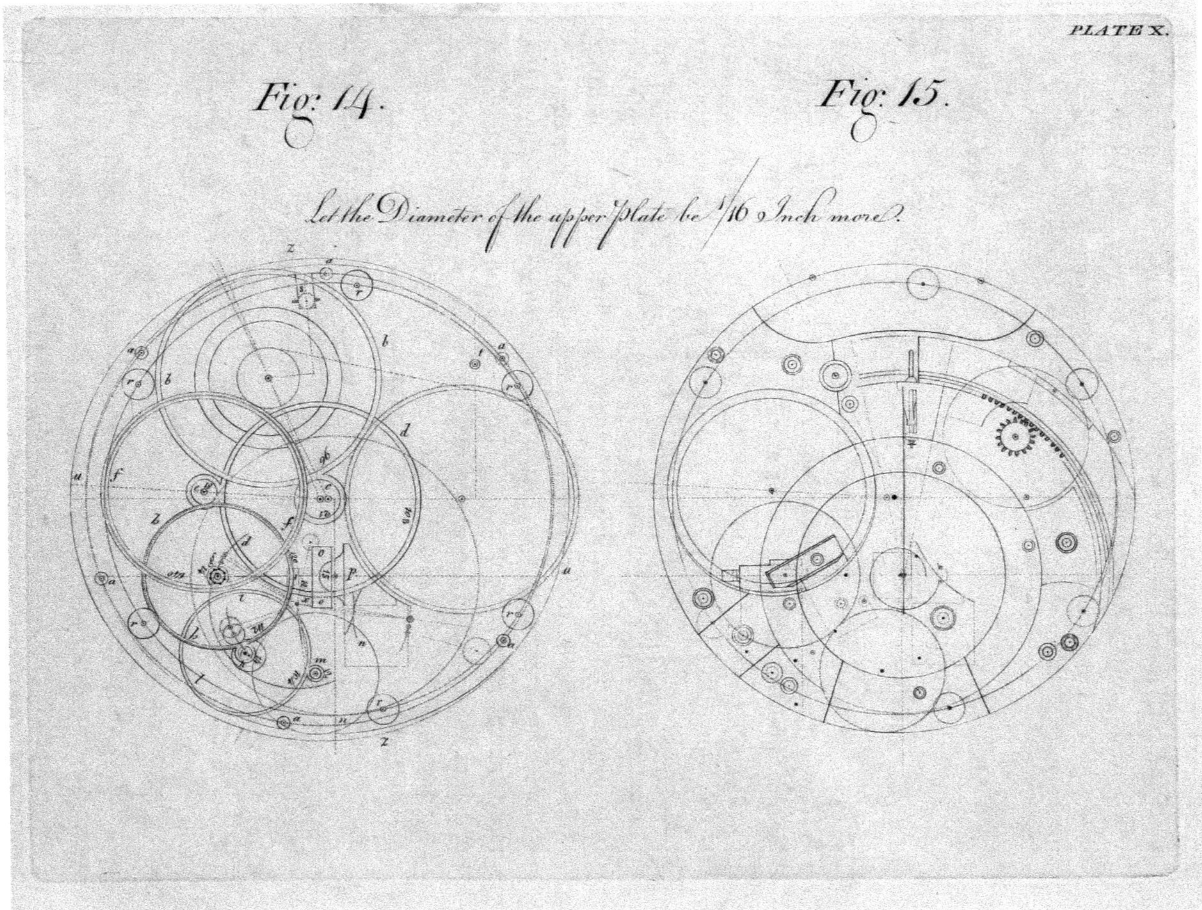

Figure 38: Plate X of *The Principles of Mr. Harrison's Time-Keeper* (London, 1767); engraving, 240 x 280mm, Royal Museums Greenwich, London.

then link these to the written description, and then, finally, read Maskelyne's more anecdotal notes. It is the engraved lines of diagram that give the text as a whole validity, which Maskelyne's concern to get these accurately transferred onto paper so clearly shows. Without these engraved plates, the clock could not be understood, or replicated.

'MR. BIRD ... DOING ALL HIS WORK WITH HIS OWN HANDS FOR THE SAKE OF GREATER TRUTH AND EXACTNESS'

ONE OF THE WORKMEN who attended Harrison's disclosure in 1765 was the instrument maker John Bird. His reflecting telescopes had been used in the trial of Harrison's watch to Barbados in 1763–64 and had been supported by the Commissioners as sufficiently accurate in response to the Harrisons' proposal of substituting telescopes by another instrument maker.[83] Bird was known for being a particularly exact maker, whose instruments could be relied on for observations. He made a number of instruments both for the Board and for the Royal Observatory. His particular skill was in dividing the scales on instruments, an engraving practice that we have already seen to be problematic when transferred to illustrations. The American natural philosopher and politician Benjamin Franklin (1706–90) discussed the unusual faith placed in Bird in a letter to his friend John Winthrop (1714/15–79) in 1768, commenting how 'Mr. Bird ... doing all his work with his own hands for the sake of greater truth and exactness ... is so singularly eminent in his way, that the commissioners of longitude have lately given him £500 merely to discover and make public his method of dividing instruments'.[84] Franklin thus outlined, with a tone of surprise, how the Commissioners were involved in paying Bird for the disclosure of his method, which would be illustrated and published, in exactly the same period as Harrison's *Principles*. The two publications that the Board produced of Bird's work—*The Method of Dividing Astronomical Instruments* and *The Method of Constructing Mural Quadrants*—in fact present a striking comparison with Harrison's. They show similar concerns by the Board, but also a different process by Bird of putting his lines onto paper, and greater trust in these by the Commissioners. Bird's known skill in engraving lines on metal instruments made his engraved lines on paper automatically more trustworthy.

83 The Harrisons tested Bird's telescopes in comparison to one by James Short at the Royal Observatory in August 1763, see *HJ*, pp.60–2.

84 Franklin, *Memoirs*, iii, Franklin to John Winthrop, 2 July 1768, p.370.

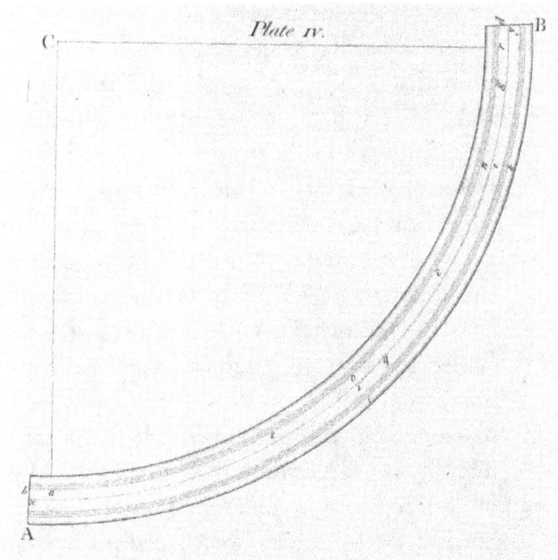

Figure 39: Plate IV of John Bird, *The method of dividing astronomical instruments* (London, 1767); engraving, 234 x 210mm, reproduced by kind permission of the Syndics of Cambridge University Library [Hh.13.9].

Bird's two *Methods* work together in the same format that the *Principles* had established, using four engraved plates (Figures 39–42). Each has a 'Preface' by Maskelyne, then one by Bird, followed by the main text. All four plates are prominently marked 'J. Bird sculpt', however, making a point for the fineness and accuracy of Bird's engraved lines, in these copperplates as well as on his instruments, before the descriptive text is even reached. Maskelyne comments in his 'Preface' on the wishes of the Board in making this publication, describing how they wanted 'complete engraved Plates of the ... Instruments, so minutely exact, as that these Plates, with the ... Description, and Models thereof, if necessary, may enable any diligent and experienced Workman to make and execute such Instruments'.[85] We notice how their intentions have become more explicit since Harrison's *Principles*. In Bird's own preface to the *Method of Dividing* he further addressed the question of the authorship of the text, explaining that he owed some of the method to his previous master, the instrument maker Jonathan Sisson (1690–1747), but 'what I call my own, I have distinguished by Italic characters'.[86] He specifically used his style of inscription to differentiate his text from Sisson's in a similar manner to how Halley or Mountaine and Dodson distinguished their definitive from their suggested lines of magnetic variation. Each used different visual techniques to identify types of information. He also made his authorship and claim to rights over his ideas very plain. We note in that context that both publications

85 Bird, *Method of dividing*, pp.iii–iv.

86 Bird, *Method of dividing*, p.v.

are explicitly authored *by* Bird on the title page, whereas the *Principles* is merely *about* Harrison's timekeeper.

In discussing his *Method of Dividing*, Bird used the fourth, simplest and smallest, plate in the set (Figure 39), which represents the scale as the only realistically illustrated part of an otherwise diagrammatic quadrant. He explained, 'it will be sufficient, for my purpose, to make use of no more Lines and Arcs than are represented in the Figure C B A, Plate 4'.[87] We see how he partly solved the problems of scale that Jacob Rowe had encountered in 1725. Bird further emphasised how very fine the engraved lines on the quadrant needed to be to create an accurate scale, and how skilled the engraver's hands needed to be in the process, saying, 'if the said impression be not too faint, feeling, as well as seeing, will greatly contribute to make the points properly'.[88] He frequently addressed the reader of the text directly in the assumption that this would be a fellow workman in the act of dividing a scale. Such instructions are all in the italic font marking out his own contribution. *The Method of Constructing* appeared a year later with the previous three plates. Maskelyne explained at the end of an otherwise largely identical preface that Bird had 'since delivered to the Commissioners three other plates … with a written paper … which forms one work with the Method of Dividing'.[89] Bird's text here worked like the *Principles*, explaining how the illustrated small parts reference into the larger figures, but we note that the most complex image of the full quadrant is shown first, followed by a simplified version and the plates of small parts (Figures 40–42).[90] Bird said confidently of the second figure after a one-sentence explanation, 'this I think is so obvious from a sight of the figure only, as to need no farther description'.[91] He explained clearly the rationale behind creating the different plates and especially the small part illustrations, saying, 'I presume it will be much more easy for the reader to understand Fig. 1. to be furnished with all the apparatus described in Fig. 2d and 3d, than if the whole had been crowded into one figure' (Figure 42).[92] Although these are diagrammatic technical drawings, Bird also used different types of line to show depth in the instrument, explaining, 'the several places in the figure mark'd by astericks (*) shew where a screw-head lies under the perpendicular bar … All the dotted lines, represent the edges of the several parts which

87 Bird, *Method of dividing*, p.1.
88 Bird, *Method of dividing*, pp.4–5, quotation p.5.
89 Bird, *Method of constructing*, pp.v–vi.
90 Bird, *Method of constructing*, pp.19–23.
91 Bird, *Method of constructing*, p.15.
92 Bird, *Method of constructing*, p.17.

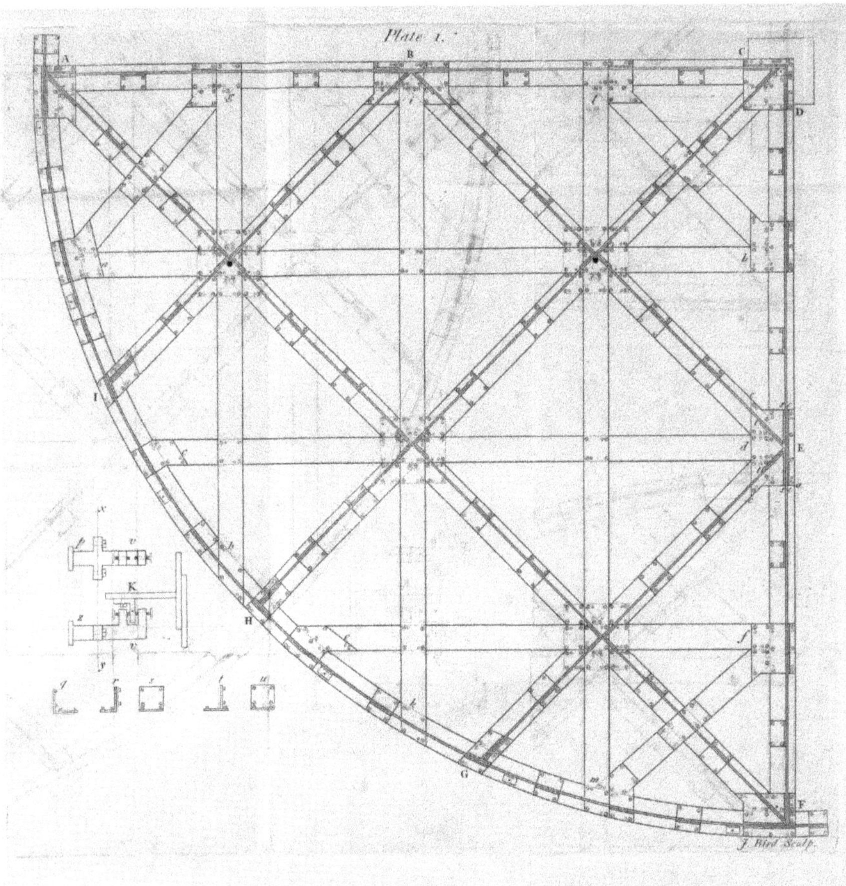

Figure 40: Plate I of John Bird, *The method of constructing mural quadrants* (London, 1768); engraving, 516 x 366mm, reproduced by kind permission of the Syndics of Cambridge University Library [Hh.13.9].

lie under, and are covered by others'.[93] He also addressed the issue of scale, in wanting to provide a suitably accurate image, but also wanting to keep relevance between the figures. He described on Plate II how, 'the figures *gg*, and *hh*, represent the iron plates, and pins, upon which the Quadrant is suspended ... The Figures *AA*, are two sections of the frame ... I have drawn these figures from a large scale, that their parts might be the better expressed'.[94] A 'scale of feet' at the bottom of the plate makes these different sizes clear (Figure 41).

93 Bird, *Method of constructing*, p.9.
94 Bird, *Method of constructing*, p.14.

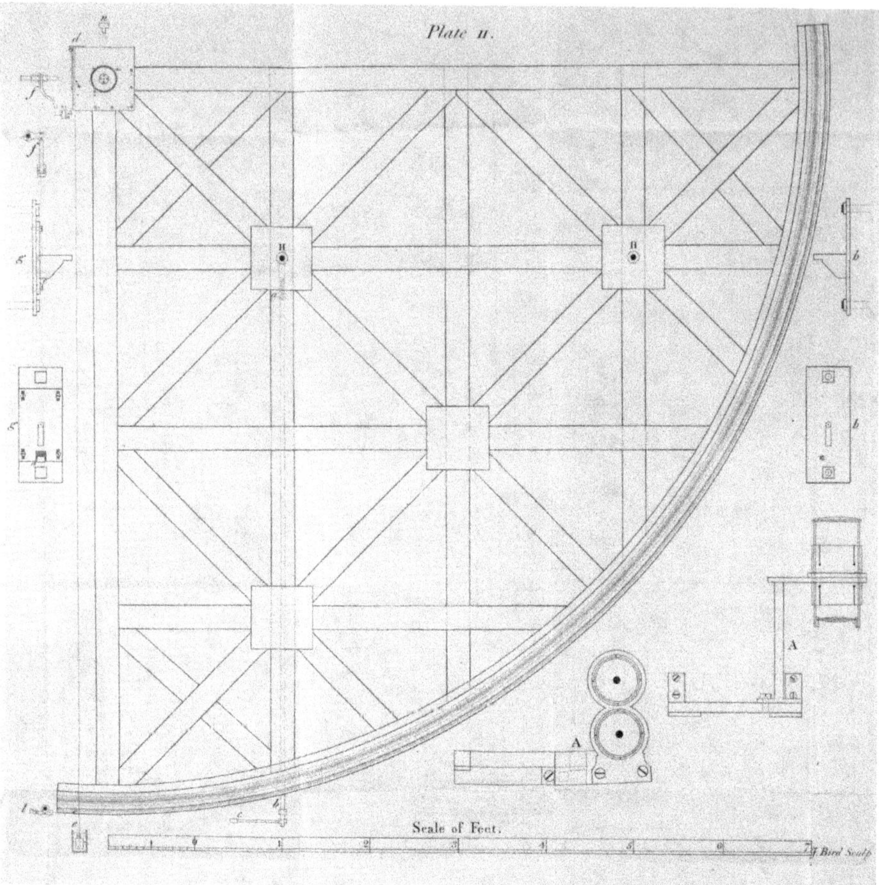

Figure 41: Plate II of John Bird, *The method of constructing mural quadrants* (London, 1768); engraving, 522 x 382mm, reproduced by kind permission of the Syndics of Cambridge University Library [Hh.13.9].

Bird, then, was confident in the creation and use of such lines, not only on paper, but on the engraved plate that transferred them there. His skill at dividing metal instruments transferred to creating their paper records. If we look at the minutes of the Board of Longitude meeting where the publication was agreed—the same meeting at which the plates of Harrison's *Principles* were sent to the watchmaker Larcum Kendall (1719–90) to begin work on a copy of the watch—we note that Bird approached the Board himself to propose the publication of 'an Account of his said Method in writing',[95] but that it was the Commissioners who added the requirement for 'Complete

95 Meeting of the Board of Longitude, 26 April 1766, RGO 14/5: 128.

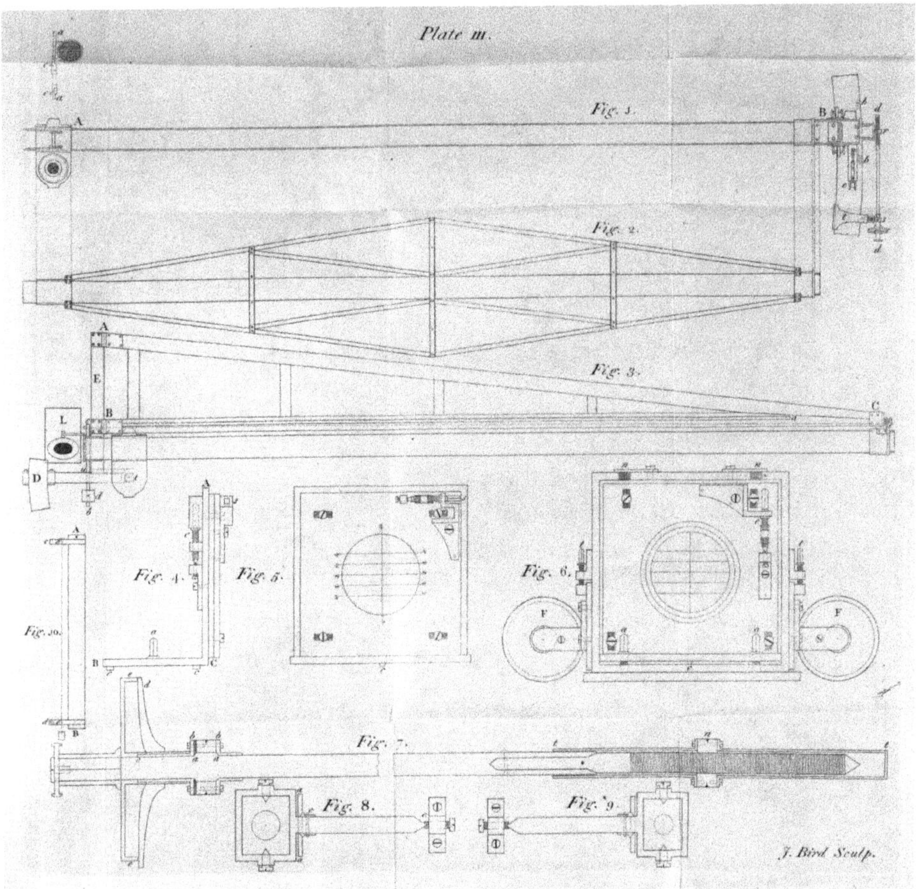

Figure 42: Plate III of John Bird, *The method of constructing mural quadrants* (London, 1768); engraving, 478 x 381mm, reproduced by kind permission of the Syndics of Cambridge University Library [Hh.13.9].

engraved Plates of such Instruments'.[96] We note also that Bird was paid thirty pounds for these plates, on top of his five hundred pound award; a payment that Harrison particularly complained had not been extended to himself and his son. Clearly the Board had profited from their negotiations with Harrison over the publication of his *Principles*, learning to converse with images in the way that this chapter has argued so many other longitude protagonists did. But Bird was also clearly more confident in his creation of lines than Harrison.

Both men had their portraits painted and engraved in these decades (Figures 43 and 44). They appear very similar: seated, neatly dressed and

96 Meeting of the Board of Longitude, 26 April 1766, RGO 14/5: 130.

Figure 43: Valentine Green, after C. Lewis, *John Bird of London* (1776); mezzotint, 426 x 320mm, Royal Museums Greenwich, London.

bewigged in the same manner, each accompanied by a sample of their mechanical skill, and each gesturing to an engraved sheet on the table at their elbow. But if we examine these sheets, we see that Bird (Figure 43) gestures at what is recognisably Plate I from *The Method of Constructing Mural Quadrants* (Figure 41), whereas Harrison's engraving appears to be merely a generic

Figure 44: P. L. Tassaert, after Thomas King, *John Harrison* (1767); mezzotint, 486 x 358mm, Royal Museums Greenwich, London.

image of a wheel; it does not match any of the plates from the *Principles* (Figure 44). Likewise, Bird's published text is tucked under his elbow, with the title clearly legible, and the inscription beneath placing it in the context of his work for the Board. Harrison's text is nowhere to be seen.

Conclusion to Part I

IN 1766, HOGARTH'S WIDOW, Jane (c.1709–89), applied to Parliament for an amendment to the Engravers' Act, which would grant her a further twenty years' copyright in her husband's works. The resulting act also extended copyright to all engravings, including works published in books, and specifically included maps.[1] Significantly, therefore, as Harrison's *Principles* was published on 16 March, and the Act did not come into effect until 6 June, his engraved plates were not covered by this copyright, compounding Harrison's complaints. Yet these changes in copyrighted images helped to re-ignite the unresolved issues around literary copyright that had smouldered between English and Scottish booksellers since the 1730s. In the resulting debates in 1769, the judges for the Court of King's Bench not only referenced the recent Engravers' Act but also entwined the question of authorship with questions of patents and mechanical invention, asking in what ways books and machines, authors and inventors were alike.[2] Harrison became the reference point in these debates. Lord Justice Yates (1722–70) put the question in precisely the terms used by Harrison, arguing that

> We have recent Instance, in Mr. Harrison's *Time-piece*; which is said to have cost Him twenty Years Application ... might not he insist upon ... the same Foundation of moral Right, for Property in *his* Invention, as an Author can for *His*? If the Public should rival Him in his Invention, as soon as it comes

1 Act 7 Geo.3 c.38, Parliamentary Statutes online. Thus, Samuel Dunn's 1775 map of magnetic variation is 'published according to Act of Parliament'.

2 Johns, *Piracy*, p.124.

> out, might not He *as well* exclaim, as an Author, 'that They have *robbed* Him of his Production, and have *iniquitously reaped* where They have *not sown*?'[3]

In a second debate in 1774, Baron James Eyre (1734–99) further pointed out how it was only the difference between inscription on paper as opposed to metal instrument that currently divided the different copyrights, saying,

> Mr. *Harrison*, after constructing a Time-Piece … hath no Method of securing an exclusive Property in that Invention, unless by a Grant from the State; yet if he was in a few Hours to write a Pamphlet, describing the Properties, the Utility and Construction of his Time-Piece, in such Pamphlet he would have a Right secured by Common Law, though the Pamphlet contained exactly the same ideas on Paper, that the Time-Piece did in Clock-Work.[4]

One of the other justices involved in these debates was Lord Mansfield who, we have seen, would rule four years later that a patent could be granted only with a fully drawn specification that disclosed its mechanism.[5] Thus, the different efforts of Harrison and Hogarth would culminate both in the establishment of literary copyright in the author, and in the basing of patent rights on the drawn image on the page. In raising questions around the production, purpose, and security of words and images on the page, the search for longitude articulates for us a particular group of visual problems that concerned both artists and scientists in eighteenth-century London.

3 *Question concerning Literary Property*, p.70.

4 Donaldson, *Cases of the appellants*, pp.33–4.

5 Johns, *Piracy*, pp.140–1.

II

The Mental Problem

'if the Love of Money cannot be said to be the Case, they must be no better than as if out of their Senses'

So says a weather-beaten—worn out Tar
Who now unfit for sea—lays up in Port
And trys with Reason's tackle—Winch or Bar
To draw fair Truth from the [Profane] of Thought.

Then beam a smile benign on the old lad:
And when you hear Brain-sweat runs through his cap
While spreading the Earth's surface on a plane
Or fathom, off his lines for a true Map

Do not cry out—such labour is in vain
Nor fancy he is Foolish—Drunk—or Mad.[1]

1 James Straycock to the Board of Longitude, 26 August 1824, RGO 14/51: 182v.

3

Proposing or Projecting?

HOGARTH'S ENGRAVING is a rare visual satire on longitude amongst the maps, diagrams, and instrument illustrations. But Hogarth himself was rare in the depth of detailed satire that he brought to his images of eighteenth-century life. Where his approach fits much more obviously is with the literary satire of his contemporaries. Bedlam, as the home of madness, was a focus of discussion and curiosity throughout our period, used by writers and artists alike to think about man, society, and the status of knowledge claims. Certain generic types appeared repeatedly within the madhouse, just as they do in Hogarth: monarchs, politicians, religious enthusiasts, philosophers, melancholic lovers, musicians, tailors. Works representing madness and madhouses, from Jonathan Swift to Hildebrand Jacob (1692/3–1739), recycled, referenced, and satirised these stereotypes, as well as their long history of visual iconography.[2] Many contemporaries also placed longitude in this context, raising questions about the authority and value of innovation entwined with concerns about language use and about how to measure the credit and credibility of new ideas. Considering these discussions further illuminates the battles that John Harrison had communicating his discoveries to the Commissioners of Longitude. It was not just what he drew, but what he said, and how he said it, which made his knowledge acceptable or dangerous to the Board.

2 Jacob, 'Bedlam', pp.11–25 and anything by Swift. MacGregor has discussed Hogarth as creating portraits of specific psychiatric disorders in *Discovery of the Art of the Insane*, pp.11–24. Wittkower and Wittkower have discussed the special place of the artist between genius and madness in *Born under Saturn*, pp.98–132.

'A MUCH HIGHER NAME THAN THAT OF A PROJECT, IF OUR LANGUAGE AFFORDED ANY SUCH TERM'

BY THE MID-1730S, hopeful pamphleteers had been proposing solutions to the longitude problem for twenty years. Yet there was no obvious sign that the Commissioners appointed by the 1714 Act, to consider such proposals and allocate the coveted reward money, had even met or given serious consideration to any of these attempts. Hopeful contributors, therefore, began to complain vocally about this lack of response, and to accuse the Commissioners of inattention to their responsibilities. In 1732, Robert Browne of Wapping (fl.1728–32), who had by this point been acquainted with Flamsteed for a number of years, published his fourth attempt since 1714 to get the Commissioners' attention, appealing directly to Parliament and criticising their inactivity:

> your Complainant has made all possible and repeated Application to the Admiralty, Royal Society, and the Professors of *Oxford*, &c. but can obtain no Relief by Reason of unjust Prejudice carried on against him ... therefore humbly hopes that for a Publick Good his Case may be taken into Publick Consideration, and a Committee appointed to examine into the Merits of this Affair ... or that the Act may be repealed to prevent the Ruin of others expecting Benefit by it.[3]

He made brutally clear how much the reaction to the Act had been initiated by the reward money and added his name to a long series of contributors who opined that working on the longitude problem had ruined them. By the point at which Hogarth was working on *A Rake's Progress*, the Longitude Act had been in operation for over twenty years, many such proposals had been published, and yet no solution had been officially tested, and the Commissioners seemed uninterested. Longitude had, in fact, become an infamous 'project', discussed across the whole breadth of contemporary literature as a scheme that distracted men's minds and brought no reward. Projects could ruin investors, as well as inventors, and were part of wider concerns about the rise of financial 'credit' and its instability. By the time of Johnson's *Dictionary*, the very definitions of longitude and 'project' became entangled. We have already seen how 'Longitude' was defined by Arbuthnot's reference to Whiston's proposal, 'his was the method of discovering the *longitude* by bomb vessels',[4] a quotation that is taken from the satirical *Memoirs of Martin Scriblerus*. 'Project' was, conversely, defined by Johnson through referencing

3 Browne, *To the Honourable*, title page.

4 Johnson, *Dictionary*, ii, 4D4v.

Joseph Addison's (1672–1719) comment in *The Guardian* that Whiston and Ditton's proposal 'is discovering the longitude, and deserves a much higher name than that of a *project*'.[5] The complex relationship of longitude to the idea of projecting, which would allow Johnson to use these two definitions by 1755, is what this chapter considers.

'Projecting' was of deep concern to many contemporary writers. It represented the imposition on the public of new, naïve, ridiculous, impossible, or even malicious schemes, the perversion of established institutions and knowledge by untested inventions. The 1707 broadsheet *The Ass Age* represented a projector among the figures ruining contemporary society (Figure 45), and Daniel Defoe (1660–1731) highlighted the complex nature of 'projects' at the end of the previous century in his *Essay upon Projects*. He nicely epitomised the problems associated with 'such a multitude of Projectors ... who besides the Innumerable Conceptions which [die] in the bringing forth, and (like Abortions of the Brain) only come into the Air, and dissolve, do really every day produce new Contrivances, Engines, and Projects to get Money'.[6] He thus distinguished between honest and dishonest projectors, all of whom claimed to improve human life, but some of whom did so out of the sole desire of making money, and cared little whom they ruined along the way: they 'rais'd people to such height, that merely on the shadow of Expectation, they have form'd Companies ... and when the Inventors have carri'd on the Jest ... they leave the Cloud to vanish of it self'.[7] Such fears of misled investment peaked at precisely the point when patenting increased in popularity, leading to the stipulation that no patent could be 'authored' by more than five people.[8] Patents were, therefore, potentially also projects. Importantly, 'projectors' could be those in charge as well as hopeful pamphleteers;[9] therefore, not only their applicants but the Commissioners of Longitude themselves could be guilty of 'projecting'. These ideas applied to changes, or proposed changes, in politics, religion, finance, literature, and even art, as well as to projects in natural and mathematical philosophy. The sheer weight of examples brought to bear in the following pages shows how widely such references were used and were constantly related to longitude.

5 Johnson, *Dictionary*, ii, 5C4r. Johnson's choice of references is particularly interesting given how often he borrowed scientific definitions from Bailey. See McCracken, 'The Drudgery of Defining' in McDermott, *Ashgate Critical Essays*, pp.73–6.

6 Defoe, *Essay upon Projects*, p.4.

7 Defoe, *Essay upon Projects*, pp.11–12.

8 Macleod, *Inventing the Industrial Revolution*, p.55.

9 Yamamoto discusses how Tory and Whig concerns about finance, religion, and politics interacted with ideas on projecting in 'Piety, Profit and Public Service'.

Figure 45: Anon., *The Ass Age; or the World in Hieroglyphick* (1707); letterpress and engraving, 472 x 337mm, © Trustees of the British Museum; and detail.

Defoe's discussion shows how, from the start, projecting schemes were associated with mental disorder—'Abortions of the Brain'—and with ephemeral bubbles or clouds, which 'only come into the Air, and dissolve'. Proposals were seen as over-inflated, ephemeral schemes that burst when touched. 'Bubbles', 'bubblers', and 'bubble-men' all, therefore, became common linguistic expressions. Bubbles became visual tropes in broadsheets criticising contemporary projects like *The Bubblers' Mirrour* of 1720.[10] The language of both these projects, and of their critics, was itself a 'bubble', using over-inflated claims for success as well as exaggerated criticisms of others, and existing within the new, rapidly expanding ephemeral print culture. This was, further, intimately connected to ideas and imagery associated with changing notions of credit, credibility, and credulity. As new systems of financial credit grew in conjunction with new systems of patronage and pensionary politics, as well as new ideas about the power of words and experience in religious belief, credit became entwined with concerns about what systems of knowledge could be relied on and believed in. Defoe constructed his worlds of good and bad credit by linking the good with natural philosophy and trade, the bad with stock-jobbers and the devil.[11] This division was achieved through questions of language use and how this could make claims credible.[12] Such imagery stretched from the seventeenth century through our period, pre-dating the infamous South Sea Bubble (of which more later), and informing discussions of any new 'project' in society. It was used by every sort of writer from the polite Addison to the scurrilous Ned Ward (1667–1731) and provides a discourse linking across genres to discuss questions of innovation in the period, from which longitude emerges as the most (in)famous example.

The Longitude Act coincided in 1714 with the Schism Bill and would soon be followed by the death of Queen Anne (1665–1714) and the fall of the Tory ministry. Thus, the longitude problem was publicised at a time when projects of a grander political and religious nature were also attracting attention and was discussed in precisely the same venues from court to coffee house.[13] Right from the passage of the Act, discussions of longitude negotiated the problematic nature of the word 'project'. In his 1714 pamphlet, *An Essay for the Discovery of the Longitude at Sea*, Flamsteed's 'non Conformist preacher from Worksop' Isaac Hawkins (fl.1714) seemed to use the term unproblematically, commenting that he knew his scheme was imperfect but wanted to share it with the public 'rather

10 Anon., *The Bubblers Mirrour* (1720); mezzotint, 390 x 290mm, Lewis Walpole Library.

11 Schaffer, 'Defoe's Natural Philosophy', pp.13–44.

12 Poovey, *History of the Modern Fact*, pp.158–68.

13 Stewart discusses this context in 'The Longitudinarians' in *Rise of Public Science*, pp.183–211.

than conceal one distant Hint from a better Projector'.[14] In their broadsheet explaining their *Reasons for a Bill proposing a Reward for the Discovery of the Longitude*, Whiston and Ditton similarly used the word as a synonym for proposal, explaining that 'this Bill is unexceptionable because it is general, and not confin'd to any one Project, Person, or Method',[15] and in the process of discussing the bill in Parliament, Roger Cotes (1682–1716) commented that, 'the project was right in the theory near the shore'.[16] Yet, in his *Guardian* advertisement referenced by Johnson, Addison recognised the dangerous connotations of the word. In Chapter 1, we saw the arguments between Henry Wilson and Thomas Haselden over their different map projections. In his heavily sarcastic critique of Wilson, Haselden associated Wilson with just such malicious projecting, commenting, 'Tis pitty, Sir, That you were not better known in the late Bubbling Times'.[17] Thus, whether seriously or satirically, longitude projects presented a potentially ephemeral chance of success.

This attitude dealt with two main concerns that accompanied longitude solutions, both of which were brought into focus by the large reward money offered by the Longitude Act: the dangers of exaggerated advertising and of time wasting. Indeed, some complained that the reward distracted interest from other more important or worthy schemes. Stephen Hales, who proposed a means of better ventilating ships in 1743, commented, 'I believe a much less Reward than what is offered for the Discovery of the Longitude, would spur People on to engage in a Research; which … is of vastly more Importance to Navigation, than the … Longitude'.[18] Thus, one concern was that proposers were impoverished inventors trying to sell their machines to the public, who might, therefore, either tack a potential use for longitude onto the description of an instrument that they were trying to sell for general navigation, or make more generally exaggerated claims of its benefits. Many longitude proposals were effectively extended adverts, rather than technical explanations. In seeking investment, such proposals promised future gain, and therefore sought to communicate in a style that investors would understand. Thus, in 1714, William Derham trod a fine line in his work on practical clock-making, saying at one point, 'I hope the more learned Reader will excuse my using many words … and that I have condescended to low things (and to him needless) as teaching the Golden Rule'[19] but at others 'If I have at any Time invaded the

14 Hawkins, *Essay*, p.5. Flamsteed used this description in a letter to Sharp. Flamsteed, *Correspondence*, Letter 1,360, p.700.

15 Whiston, *Reasons*.

16 *Journal of the House of Commons* (11 June 1714), pp.677–8.

17 Haselden, *Reply*, p.25.

18 Hales, *Description*, p.163.

19 Derham, *Artificial Clock-Maker*, pp.5–6.

Workman's province, it was not because I pretended to teach him his Trade; but ... for Gentlemen's sakes'.[20] He sought to appeal both to the artisan who might copy his work, and to the gentleman who might invest in it.

The second concern was the idea that longitude was in fact an insoluble problem, and therefore a waste of time and money for inventors and investors alike. So, the very nature of the longitude problem could be interpreted as showing that only the naïve projector would attempt to solve it. Even the society correspondent Horace Walpole joked about the projecting nature of longitude schemes in a letter to Richard Bentley (1708–82) in the 1750s, saying, 'I cannot support a commerce of visions! and therefore, whenever you send me mighty cheap schemes for finding out longitudes ... you will excuse me if I only smile, and don't order them to be examined by my council. —For heaven's sake, don't be a projector!'[21] In 1752, the editor of the *Gentleman's Magazine*, Edward Cave (1691–1754), found it necessary to publish a notice 'To the Gentlemen who have sent us Proposals on the Longitude', commenting that the search for longitude was a worthy cause, so they had always offered space to it, but 'think it necessary to signify that many schemes have been sent us, which to publish would do no honour to their authors nor service to the community ... some for effecting utter impossibility ... The authors of all such will not, 'tis hoped, blame us for suppressing their papers'.[22] Such impossible schemes clearly existed within the world of advertising and commercialised language. Indeed, some longitude pamphlets recognised this dangerous view of their subject and felt compelled to argue that previous lack of success did not prove its impossibility. In 1735, when the mathematicians William Ward (fl.1735) and Caleb Smith (fl.1735–40) contributed their description of a new quadrant to the debates in the previous chapter, they commented perceptively that

> The various *Idle* Schemes and *Chimerical* Projects that have been offered as *Discoveries* of the *Longitude*, have so much prejudiced Men's Minds against all Propositions of this sort, and brought so much Disgrace on the Projectors, that every *Attempt* to solve this *valuable Problem*, is now ridiculed as the effect of a weak, or a distempered Brain: The Thing itself has been so long the *Reproach of Art*, and every *Essay*, how promising soever, has hitherto proved so *unsuccessful*, that the greatest part of Mankind look on it as an *Impossibility*.[23]

Interestingly, the contemporary discussion of this proposal in the review journal *The Present State of the Republick of Letters* paid much greater attention

20 Derham, *Artificial Clock-Maker*, pp.8–9.

21 Walpole, *Correspondence*, xxxv, pp.190–1.

22 *Gentleman's Magazine* 22 (August 1752), p.359.

23 Ward, *Description and use*, p.4.

to the concern of impossibility than to their proposed instrument.[24] Commercialised literary productions were clearly not immune to concerns about their own credit and credibility.

The Projector in Bedlam

Bedlam was particularly popular with authors as a setting for discussions of contemporary society, and often featured representations of the natural philosophical or mathematical projector as a stereotypical inmate. In 1727, an English translation appeared of *Respublica Literaria*, the satire on learning produced by the celebrated Spanish man of letters Diego de Saavedra Fajardo (1584–1648) in 1670. In this, he had made mathematical projecting the embodiment of Bedlam madness and referenced longitude specifically in the activities of the porters on the door to the hospital. These men were

> each in a brown Study aiming at Impossibilities, in a manner, and heedless of all that came and went. The one with a half-starved and drowzy Countenance, was endeavouring to square the Circle upon the Wall: The other flatter'd himself with having made an Instrument for finding out the Longitude.[25]

Interestingly, the original Spanish describes the second porter as flattering himself 'with more greed than glory', the very model of a projector, and the first as squaring the circle 'with a compass in his hand, on a blackboard' in just the manner of Hogarth's longitude lunatic.[26] This also has obvious visual connections to Caius Gabriel Cibber's (1630–1700) statues that crowned the gates of Bedlam in Moorfields (Figure 46). We note that the porters' activities are defined as impossible, as well as involving newly invented instruments.

Longitude projecting, as an impossible and commercialised activity, was thus making dangerous inroads into the republic of letters, the idealised communication network that linked learned men in the period through agreed forms of credible and creditable language use.[27] Defoe, likewise,

---

24 *Present State of the Republick of Letters*, xv, pp.75–9.

25 Saavedra, *Respublica Literaria*, pp.112–13.

26 'Mas atentos à veneer lo cassi impossible de sus empresas, que à los que entravan, y salian: El uno macilento, y desvelado; con un compass en la mano, procurava sacar sobre una pizzarra negra, a quadratura del circulo; y el otro con mas codicia, que gloria, formava un instrumento Mathematico, con que se persuadia haver hallado en la navegacion la certeza de la longitud', 'Republica Literaria', i, p.20. My thanks to José Ramón Marcaida for this translation.

27 As Schaffer has argued in 'Defoe's Natural Philosophy'.

Figure 46: Andrew Birrell, *Figures over the Gateway of Bethlehem Hospital* (1813); engraving, 133 x 207mm, Wellcome Collection, Public Domain.

had referred to projectors as 'authors', making a clear relationship between commercialised natural philosophy and commercialised literature,[28] and it was the problematic relationship of commercialised language to creditable knowledge that concerned the Scriblerians. Swift, Pope, and Arbuthnot particularly linked scientific projecting to 'projects' in all other branches of society, to politics, religion, finance, and especially literature. They emphasised that 'bad' projects could be identified by their 'bad' language. In *A Tale of a Tub*, Swift directly equated scientific projecting and Bethlehem madness, saying of 'philosophers' that

> several of the chief among them, both *ancient* and *modern*, were usually mistaken by their adversaries, and indeed by all except their followers, to have been persons crazed, or out of their wits ... agreeing for the most part in their several models with their present undoubted successors in the *academy* of *modern Bedlam*.[29]

28 This has been discussed both by Schaffer in 'Defoe's Natural Philosophy', pp.13–44, and by Ratcliff, 'Art to Cheat the Common-Weale', pp.337–65.

29 Swift, 'Tale of a Tub' in *Major Works*, p.141.

Figure 47: Bernard Lens and John Sturt, Figure 6 from Jonathan Swift, *A Tale of a Tub* (London, 1743); etching, 173 x 102mm, reproduced by kind permission of the Syndics of Cambridge University Library [7720.d.1134].

Indeed, an engraving accompanying early copies of *A Tale* featured a representation of Bedlam, designed by Bernard Lens the Elder (1659–1725) and engraved by John Sturt (1658–1730), which bears startling similarities to Hogarth's engraving, including an inmate gesturing at a drawing on the wall (Figure 47).[30] Swift took this idea to its logical conclusion in *Gulliver's Travels* where in the Academy of Lagado—part Bedlam, part Royal Society,

30 MacGregor has suggested that this may have inspired Hogarth. See *Discovery of the Art of the Insane*, pp.16–17.

part Exchange Alley, part Grub Street—the philosophers are found in rooms that resemble cells, pouring over crack-brained schemes.[31]

In his 1736 poem 'The Character, Panegyric and Description of the Legion Club', Swift extended the same idea further, to the members of the Irish Parliament, and even linked this to Hogarth's Bedlam, commenting,

> How I want thee, hum'rous Hogarth!
> Thou, I hear, a pleasant rogue art.
> Were but you and I acquainted,
> Ev'ry monster should be painted:
> You should try your graving tools
> On this odious group of fools.[32]

Alexander Pope's poem *The Dunciad* likewise specifically related literary projecting, or hack writing, to the madness of Bedlam. He highlighted how the seat of literary dullness—Grub Street—was to be found next to Bedlam's Moorfields site in London, making it an easy step from one to the other:

> Close to those walls where Folly holds her throne …
> One cell there is, concealed from vulgar eye,
> The cave of poverty and poetry …
> Hence *Journals, Medleys, Merc'ries, Magazines*:
> Sepulchral lies, our holy walls to grace,
> And New Year odes, and all the Grub street race.[33]

We note that it is as much the deluge of production from Grub Street—'*Journals, Medleys, Merc'ries, Magazines*'—as it is the poor language used which worried Pope. He feared the continuing expansion of print technology, and how it allowed indifferent or even morally problematic works to be encased within printed structures that gave a veneer of credibility and credit.[34] Swift's Laputan satire took this to its extreme conclusion by creating a group of projectors who used a machine to distil universal knowledge into new books, whereby 'the most ignorant Person at a reasonable Charge, and with a little bodily Labour, may write Books in Philosophy, Poetry, Politicks, Law, Mathematicks and Theology, without the least Assistance from Genius or Study'.[35] The rush to get into print overrode any time to reflect properly.[36]

---

31 Swift, *Gulliver's Travels*, pp.167–79.
32 Swift, 'Character, Panegyric' in *Major Works*, p.561.
33 Pope, 'The Dunciad' in *Selected Poetry*, p.130.
34 Kernan, *Printing Technology*, pp.9–15.
35 Swift, *Gulliver's Travels*, p.171.
36 Lynall, *Swift and Science*, p.7.

'A MULTITUDE OF TRIFLING AND FRUITLESS SCHEMES AND PROJECTS'

JOHN ARBUTHNOT, a physician and natural philosopher in his own right, played on precisely these forms of mathematical and literary projecting in his 1716 *Humble Petition*, making longitude projecting the ultimate end of both. The *Petition* purported to be from various trades, including cooks and blacksmiths, against a new scheme for harnessing sunlight to create heat. It ends with the suggestion that a tax be put on sunbeams, 'reserving the sole Right and Privilege of the *Catoptrical Cookery* to ... the Commanders and Crew of the Bomb-Vessels, under the Direction of Mr. *Whiston* for finding out the Longitude, who by reason of the Remoteness of their Nations, may be reduc'd to streights for want of Firing'.[37] He not only implicated longitude as a similarly absurd project, but referenced Whiston and Ditton's specific proposal. Mathematics was, indeed, the area of projecting most often associated with longitude. Swift's Lagadan projectors give examples of the most absurd schemes with which such projectors might engage, from building houses from the roof down, to learning mathematics by ingesting its principles written on a wafer. In gaining entry to the Academy by being presented 'as a great Admirer of Projects, and a Person of much Curiosity and easy Belief',[38] Gulliver articulated the problems of credit and credulity that Swift saw in such projecting: it sat on the boundaries, transgressing the accepted edges of mathematical knowledge.

There were three specific projects with which longitude was constantly associated, all of which were notably seen as impossible, and remain so to this day: the philosopher's stone, perpetual motion, and squaring the circle.[39] Broadsheet images like *The Bubbler's-Kingdom in the Aireal-World* showed alchemical schemes like the philosopher's stone turning gold into bubbles, the ultimate image of projecting (Figure 48). Each of the inventors is surrounded by scraps of paper advertising his projects, and the frontispiece to William Hunt's (fl.1673–1737) comic play *The Projectors*, published in 1737, shows us projectors bringing just such schemes for longitude and the philosopher's stone side by side (Figure 49). In the text, we learn that Shirtless is seeking to solve the longitude to gain a dowry for his daughters: 'I have e'en almost found out the *Longitude*, and only want much about the sum you mention, to make two or three Experiments, that would complete the

37 Arbuthnot, *To the Right Honourable*, p.2.

38 Swift, *Gulliver's Travels*, pp.167–9, quotation p.166.

39 Simon Schaffer has considered the political and financial implications of perpetual motion projects in 'The Show That Never Ends', pp.157–89.

Figure 48: John Pine, *The Bubbler's-Kingdom in the Aireal-World* (1720); etching, 242 x 341mm, source gallica.bnf.fr / Bibliothèque Nationale de France.

Figure 49: Anon., Frontispiece to William Hunt, *The projectors* (Edinburgh, 1737); engraving, 192 x 121mm, Private Collection.

Discovery'.[40] Other projectors in attendance are Air-Castle and Machine, accompanied by the authors Pamphlet and Politick, and all are as interested in swindling each other's money as in forwarding their schemes. We note from the frontispiece that Shirtless, living up to his name, is very obviously the most unkempt of the projectors, and is in the most obsequious pose before the gartered nobleman.

In *The Memoirs of Martin Scriblerus*, the Scriblerians satirised contemporary writing that they disliked by making Martin the author of a whole range of pamphlets and schemes. Their evocation of these included 'many precise and Geometrical *Quadratures* of the *Circle* … the Projects of *Perpetuum Mobiles*, *Flying Engines* … the Method of discovering the *Longitude* by *Bomb-Vessels*'.[41] They conflated satirical and real-life projects, by referencing Whiston and Ditton's 1714 proposal. Elsewhere, in a wonderful satire on projecting, an author calling himself 'AM in Hydrostat' proposed a range of benefits to be got from draining the Irish channel, which included obtaining rare fur from sea foxes, finding the old moons that had fallen into it, and 'the particular Places of the Channel wherein (most likely) the Perpetual Motion, and the Longitude, are to be grappl'd up; for you will allow that 'tis matter of great Depth to find those out'.[42] In 1771, a comic poem entitled *The Ruffle* drew a direct relationship between such projects and Grub Street literature, recommending the reader to

> Look in the papers valu'd most,
> The *London* or the *Gen'ral-Post*,
> Or, in the *Publick Advertiser*,
> You'll find each day the world grow wiser …
> What cannot art of man find out?
> *Astrologers* no longer doubt
> The *Longitude*, as yet unknown,
> Will to the world be quickly shown.
> Each ign'rant *Dunce* will have a notion,
> How to construct *Perpetual Motion*.[43]

Indeed, one proposal from Elias Pledger (fl.1731) in 1731 directly blamed this continuous satire on the reward money offered by the Longitude Act. He complained that navigation had 'receiv'd less Improvement in this Age than any other Art whatsoever … greatly owing to the extravagant Rewards that have been offered to any Person, that should solve the Grand Problem

40 Hunt, *The projectors*, pp.16, 46.
41 Kirby-Miller, *Memoirs*, pp.166–7.
42 A. M., *Thoughts of a project*, p.30.
43 Lund, *The mirrour*, pp.39–40.

of finding the Longitude at Sea. This like the Philosopher's Stone hath set many Heads to work, and hath occasion'd a Multitude of trifling and fruitless Schemes and Projects'.[44]

'Quack' medicine was another means of imposing schemes on the public with which longitude was particularly associated. In 1723, Gabriel John (fl.1723), who called himself 'a seventh Son, and Teacher of the Occult sciences in Yorkshire', satirised fellow doctor John Hancock's miracle cure for the plague, saying, 'if the Doctor has the Art to make People *immortal*, what is that to any body? There must be a time to find out the Longitude, *if ever it be done*; and some body or other must do it, *if any body does*; ay, and the Philosopher's Stone too'.[45] In suggesting that eternal life would be the only means of ever successfully finding the longitude, he echoed Swift's suggestion in *Gulliver's Travels* that it would be worth the immortal but grisly life of the Struldbruggs as 'I should then see the Discovery of the *Longitude*, the *perpetual motion*, the *Universal Medicine*, and many other great Inventions brought to the utmost Perfection'.[46]

Elusive cures for particularly feared contemporary diseases, such as syphilis and bladder stones, were also frequently equated with the hunt for longitude. Attempts to cure 'the stone' were, however, given serious consideration by the government, and give us one of the only offers of money comparable to the longitude reward. In 1738–40, Joanna Stephens (b. before 1720–74) announced that she had found a cure for the stone, petitioning Parliament for a reward of £5,000, which she received after disclosing her cure to a group of trustees.[47] The parallels with longitude here are clear, and did not go unnoticed by contemporaries. In 1774, discussing precedents for parliamentary support of innovation, the journalist William Kenrick (1725–79) commented, 'the parliamentary rewards, that have been offered and paid for the finding out the Longitude … for a nostrum for the Stone, and for several other inventions and discoveries, afford a sufficient proof that the encouragement of ingenuity, by rewarding the authors of such inventions and discoveries, is in general adjudged to be politically expedient'.[48] Kenrick was a particularly vocal critic of these sorts of projects, also attacking perpetual motion schemes.[49] Harrison too would be directly compared to Stephens in 1767, as we will see.[50]

44 Pledger, *Brief Description*, p.i.
45 John, *Flagellum*, p.15.
46 Swift, *Gulliver's Travels*, p.136.
47 Viseltear, 'Joanna Stephens', pp.199–220.
48 Kenrick, *Address to the artists*, p.16.
49 Schaffer discusses the history of one such project, Orffyreus' perpetual motion wheel, in 'The Show That Never Ends', pp.159–60, 187.
50 Jenner, *Letters from Altamont*, pp.86–8.

'It may be as difficult a matter to retrieve the National Credit'

We have already discussed the problematic relationship of money and innovation, in the contemporary systems of financial credit that so concerned Defoe and the Scriblerians. The most infamous and most satirised project in the early eighteenth century was the financial and political disaster that followed the rise and fall of stock in the South Sea Company, an event known as the 'South Sea Scheme' or 'Bubble'.[51] This was just the most famous of a whole range of stock-jobbing projects in the period, of which it became emblematic by being the most spectacular.[52] In June 1720 Robert Harley, Earl of Oxford and recently disgraced Tory minister, was told by his son that 'the demon of stock-jobbing is the genius of this place. This fills all hearts, tongues and thoughts, and nothing is so like Bedlam as the present humour'.[53] He again made a direct relationship between projecting and the Bedlam madhouse. Such dangerous financial credit linked back to the concerns about projecting literature that haunted Defoe and the Scriblerians. The new financial credit systems were also, of course, based on new forms of paper money, which bore worrying similarities to the printed forms of Grub Street writing. The material credibility of both was far too ephemeral.

Many visual and literary satires dealt with events surrounding stock-jobbing and the South Sea Scheme.[54] Perhaps most appealing are the packs of 'Bubble Cards' (Figure 50), where each card shows a disastrous 'project', with a poem and small printed image. *The Bubblers' Medley* (Figure 51) further made a *trompe-l'oeil* collection of ephemeral Grub Street publications, images and playing cards, creating a commercial display of such print production itself.[55] Noticeably one of these extracts and re-arranges a poem by Jonathan Swift written in 1721, entitled *The Bubble*, in which he laid the blame for the crash firmly at the directors' door. He compared how

> … some build Castles in the Air,
> *Directors* build 'em in the Seas;

---

51 Schaffer considers Orffyreus within questions of British financial credit in 'The Show That Never Ends', pp.182–7.

52 Hoppitt, 'The Myths of the South Sea Bubble', pp.141–65.

53 Historical Manuscripts Commission, *Report on The Manuscripts*, v, p.599.

54 Hallett argues that Hogarth's early satires were particularly indebted to the Dutch examples that flourished surrounding the 1720s bubble: *Hogarth*, pp.20–4.

55 See Hallett, 'The Medley Print', pp.214–37 and Wahrman, *Mr Collier's Letter Racks* for discussions of such *trompe-l'oeil* images.

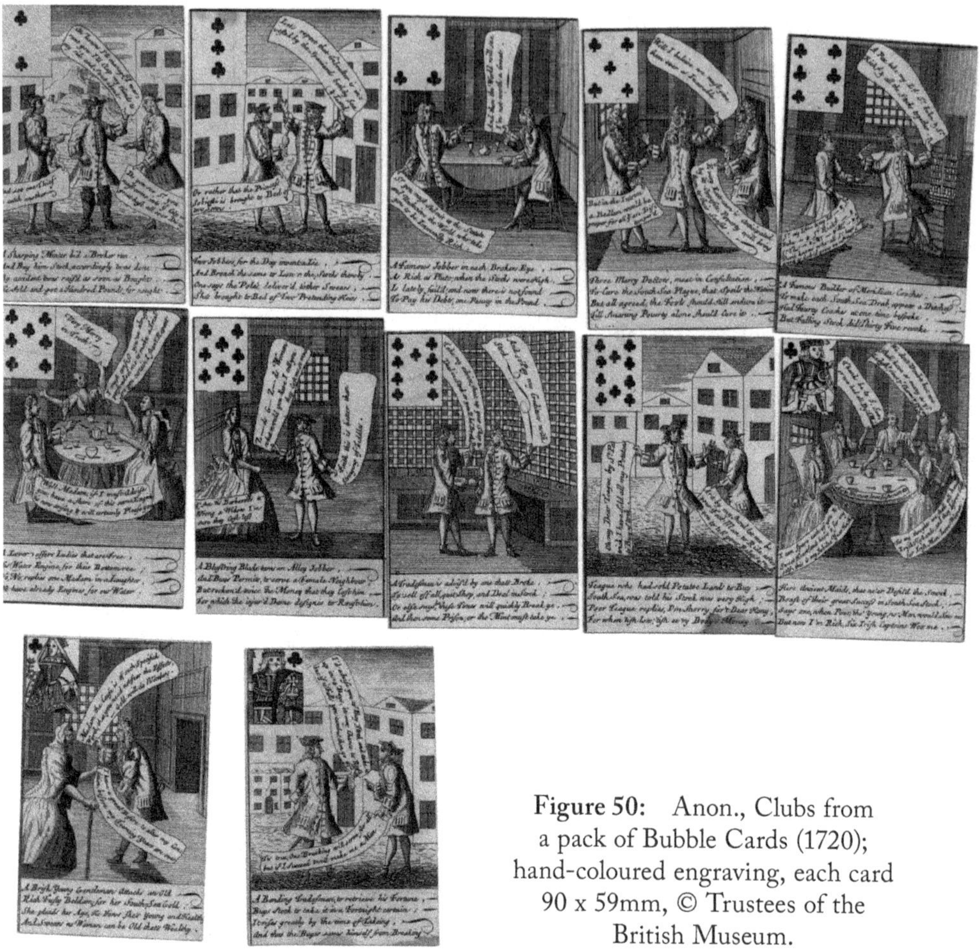

Figure 50: Anon., Clubs from a pack of Bubble Cards (1720); hand-coloured engraving, each card 90 x 59mm, © Trustees of the British Museum.

> *Subscribers* plainly see them there,
> For Fools will see as Wise-Men please.[56]

Further, he made direct relationships between financial crash and shipwreck, describing Exchange Alley as a stream in which

> Subscribers here by Thousand float,
> And jostle one another down;
> Each paddling in his leaky Boat,
> And here they fish for Gold and drown.[57]

56 Swift, *The Bubble*, p.10.
57 Swift, *The Bubble*, p.9.

Figure 51: Anon., *The Bubblers Medley* (1721); engraving, 335 x 248mm, Wellcome Collection, Public Domain.

In the so-called Drapier's Letters controversy, Swift virulently attacked William Wood's (1671–1730) scheme to introduce a new half-penny to Ireland as an example of colonial oppression. He equated its introduction with finding longitude, asking of Wood, 'he must be surely a Man of some wonderful Merit ... Hath he discovered the *Longitude* ...? No. But he hath found out the *Philosopher's Stone* after a new Manner, by *Debasing* of *Copper*, and resolving to force it upon us for *Gold*'.[58] Swift's anger at this halfpence scheme can be linked to his mistrust of Sir Isaac Newton as head of the Royal Mint. Newton's currency strategies were all part of his claims for scientific accuracy that Swift so distrusted. We note that Gulliver's travels in Laputa were written just after the Drapier's Letters conflict, and how much the Academy of Lagado could resemble Exchange Alley alongside the Royal Society.[59] Gulliver's visit to Balnibarbi was, in fact, Swift's grand satire on the financial and political projecting of the South Sea Bubble.[60] Others continued to link longitude directly with financial projects. In 1766, the mathematician and Fellow of the Royal Society Edmund Stone (1702–68) commented that longitude had become a project in the 1720s. Stone's translations had helped to make Gottfried Leibniz's (1646–16) calculus available in English, and he had produced a shorter and cheaper *New Mathematical Dictionary* to compete with Harris' *Lexicon Technicum*.[61] Within *Some Reflections on the Uncertainty of many Astronomical and Geographical Positions*, Stone described how the longitude reward had

> hitherto had no other effect than setting all heads and hands at work, and producing, for the sake of interest, abundance of quacks and impossible schemes for its discovery ... in the year 1722 the discovery of the longitude, was made one of the bubbles in *Change-Ally*, and put up to be sold to the best bidder. Many brokers, crazy projectors, and poor crafty knaves, cryed it at *Garraway's Coffee-House*, along with ... *South-sea* stock.[62]

Indeed, we have one example of a projector who did end up in the Fleet Prison for debt because of investing in his longitude schemes. In November 1733, a Samuel Clark (fl.1733) wrote to Sir Hans Sloane (1660–1753), President of the Royal Society, from the Fleet, where he was confined briefly for debt. His

58 Swift, *Some observations*, p.14.

59 Lynall, *Swift and Science*, pp.96–103. Lynall has convincingly argued that the Scriblerians were more engaged with natural philosophy than previously appreciated. See also 'Scriblerian Projections'.

60 Rogers, 'Gulliver and the Engineers' in *Eighteenth Century Encounters*, pp.11–28; Treadwell, 'Jonathan Swift', pp. 439–60.

61 Niccolò Guicciardini, 'Edmund Stone' in the *ODNB*.

62 Stone, *Some reflections*, pp.x–xi.

continuing commitment to the ethos of the republic of letters is shown by his recommendation to Sloane of 'a very ingenious young Gent. who has the misfortune to be here yet is a very good Mathematician, & from the encouragement that was given by Act of Parliament has made it his Study to find out the Longitude, & thinks he has hit upon what will do; & you being one of the Commissioners as well as President of the Royal Society, he is very desirous you should have sight of it as well for your own judgment, as for you to show it to … the Commissioners'.[63] In referencing the reward or 'encouragement' offered by the 1714 Act, Clark followed the majority of other longitude proposers in harping on the reward money in a way that we have seen so worried the Scriblerians. 'The twenty thousand pounds' is thus reminiscent of prizes in that other great eighteenth-century financial scheme: the state lottery. For many satirists there was no distinction between the longitude reward and a lottery prize. Harrison too returned constantly to the reward money, as we shall see.

Schemes like the South Sea Company and the state lottery were, of course, not merely financial projects, but also political ones. The creation of the South Sea Scheme was, after all, largely a Tory attempt to balance the Whig creation and control of two other new financial institutions: the Bank of England and the East India Company.[64] In 1766–67, the radical Wilkite, Arian, and teacher James Burgh (1714–75) looked back critically in his *Crito* at the entwined political and financial attempts that had been made to solve the national debt, commenting,

> Schemes there have been offered enough for paying the national debt, the respective authors of which have each been certain of the success of his own. I will not say, that this political problem is as difficult, or as hopeless of solution, as that of the longitude. But the fates of the two have been nearly hitherto alike: much proposed; but little executed.[65]

Swift dealt extensively with political projecting in *A Tale of a Tub*. At the end of the 'Digression on Madness', he analysed different types of madmen and which state position they would best hold if the Tories were to follow his advice 'to bring in a bill for appointing commissioners to inspect into Bedlam … by which means, duly distinguishing and adapting their talents, they might produce admirable instruments for the several offices in a state'.[66]

63 Samuel Clark to Sloane, British Library, Sloane MS 4053 f.82. My thanks to James Delbourgo for bringing this letter to my attention.

64 Hoppitt, 'Myths of the South Sea Bubble', pp.141–2. Bedlam was likewise seen as an institution whose governors supported Tory policies of control. Kromm, *Art of Frenzy*, p.103.

65 Burgh, *Crito*, i, p.33. For Burgh's political and literary career see Carla Hay in *ODNB*.

66 Swift, 'Tale of a Tub' in *Major Works*, pp.145–9, quotation p.146.

In his correspondence with Arbuthnot, Swift specifically linked political projecting to longitude, commenting on parliamentary fears of a Jacobite rising in 1714 that 'the Commons offer 100000 [pounds]: If I was the Pretender I would come over my self, and take the money to help pay my Troops. They had better put out a Proclamation that whoever discovers the Pretender or the Longitude shall have 100000'.[67] Longitude worked as a projecting joke no matter what your views. Also in 1714, Sir Richard Blackmore (1654–1729), the Tory, Anglican physician-in-ordinary and writer who would later become one of Pope's victims in *The Dunciad*, joked that he had discovered the philosopher's stone and would give it to a deserving genius but, 'If none such appear in my Life-Time, I shall leave it by my Will in the hands of Trustees, to be given to any Publick spirited Person, who shall have found out the Longitude, or shall have reconcil'd the different Parties and Interests, that divide this Nation'.[68] A year later, Daniel Burgess criticised the Tory ministry as knowing no more of politics 'than they did of the Longitude'.[69] Burgess was a pseudonym used by Thomas Gordon (c.1691–1750), the author of *Cato's Letters*, which mounted a detailed Whig critique of the Tory ministry. Thus, the longitude problem could be employed by either party as a means to accuse political opponents of 'bubbling' or malicious projecting.

'the Longitude of the Church'

One of Swift's other hated projects was the attempt to establish religious toleration. For him dissenting religious enthusiasm was a mark of national madness, as dangerous as schemes in natural philosophy. *A Tale of a Tub* dealt extensively with religion as a project, discussing, for example, 'Peter' who symbolised the Catholic church: 'after much thought he cast about at last to turn *projector* and *virtuoso*, wherein he so well succeeded that many famous discoveries, projects, and machines, which bear great vogue and practice at present in the world, are owing entirely to Lord Peter's invention'.[70] If Whigs and Tories disagreed over financial and political 'projects' they also fundamentally diverged in their views on the established church and its relationship to the state. The 'Augustan' views of the Scriblerians feared enthusiasm in both religion and natural philosophy as

67 Arbuthnot, *Correspondence*, Swift to Arbuthnot, 3 July 1714, pp.181–2.
68 Blackmore, *Lay-Monastery*, p.44.
69 Burgess, *Letter to the Bishop of Salisbury*, p.8.
70 Swift, 'Tale of a Tub' in *Major Works*, p.111.

introducing new forms of knowledge that sought to bypass divine providence in their search for truth.[71]

Longitude was, again, linked directly to religious projecting. In *A Collection of the Occasional Papers for the Year 1716*, one correspondent commented, 'The next form I find you or your correspondent in, is a Projector. Really, Sir, I have seldom observ'd any great Advantage from Expedients, and least of all do expect any from *an Expedient for Peace among all Protestants.* If you could find out such a thing, I deny not but it might be near upon as good, as a Discovery of the *Longitude*; but it seems at present … as unlikely and impracticable as that Invention'.[72] The whole longitude debate had, in fact, been tied to religious enthusiasm from the start through the person of William Whiston, who was as known for his Arian views as for his longitude schemes.[73] In *The Conventicle*, the prolific publisher and editor of the *Athenian Gazette*, John Dunton (1659–1733), linked Whiston's longitude schemes and dissenting religion, saying, '*Longitude at sea* is but just now discover'd by Mr *Whiston*, and who knows but this new discovery of *dividing the Church to unite it*, (tho' it has been long a secret to the *Anti-Schismaticks*, and much harder to find than *Longitude*) may at last prove to have as much reason and certainty in it as *Whiston's* Discovery'.[74] Chapter 4 will return to look at Whiston in more detail, but he is far from the only link between longitude and religious projecting.[75] Benjamin Parker (fl.1734–47) both preached in church and published on longitude, and used one to help gain attention for the other.[76] In his *A Survey of the Six Days Works of the Creation*, for example, he mentioned his work on longitude six times to give his religious arguments more credibility.[77] In 1767, attempts to unite Protestant churches in Britain were still being equated with the difficulty of finding longitude. *The Real Seeker* commented that 'I must own that such a vague and generical church were as useless to the unity of Christians as those undetermined points are to find out the longitude at sea, as it should seem the same thing, that there should be no church, as not to know where to find her'.[78] This, of course, takes us back to questions of accurately mapping the coast discussed in the first chapter.

71 Heyd, *Be sober and reasonable.*

72 *Collection of the Occasional Papers*, p.6.

73 Heyd, *Be sober and reasonable*, p.244.

74 Dunton, *Conventicle*, p.89.

75 Stewart, *Rise of Public Science.*

76 See, for instance, announcements in the *Daily Journal* on 8, 12, and 13 November 1734 (issues 4,309, 4,312, and 4,313) and Parker, *Parker's projection.*

77 Parker, *Survey.*

78 *The Real seeker*, p.41.

One particular pamphlet beautifully satirised the relationship between longitude projecting and projects for and against religious dissent. In 1721, Sir Richard Cocks (c.1659–1726) published a pamphlet entitled *A Perfect Discovery of the Longitude*, proposing his project for solving the problem. This is best given from his own dedicatory letter, albeit at some length, where he recounts how,

> observing in the Votes something relating to the Longitude, I remember'd formerly, that a *Premium* was offer'd to those that made the Discovery. I knew the Bishop of Bangor, and his great Opponents, were all wise Men, and concluded, that they would never quarrel for nothing. So I concluded it was about the *Premium*, for this Discovery. I know the *Latitude* has been long discover'd, but the *Longitude* was still a Secret. I have perceiv'd those great Mens Oversights, and have attain'd to a perfect Discovery of the *Longitude*. And when I had finish'd it, I shew'd my labours to a Friend, who laugh'd at me, and told me, That I mistook; for that the *Longitude*, that they meant, was for the Benefit of Navigation. I own this put a little Damp upon my Spirits, but upon Second Thoughts I concluded, that if the *Longitude* of the *Church* was not timely discover'd, our *Spiritual Pilots*, would make shipwreck of the *Nation* ... I have by my *Discovery* plac'd the *Church* in eternal Tranquility, in a safe Harbour out of all Danger, which alone is worth all the *Premium* that is offer'd for the other *Longitude*. I desire only to have the *Premium* for this, till the *Discovery* of the other is made out and approv'd of; giving Security to pay it then to the Discoverer.[79]

Cocks was responding to the Bangorian Controversy, where Benjamin Hoadly (1676–1761), the Bishop of Bangor, had argued, in a sermon preached in front of George I (1660–1727) in 1717, that the church should be entirely distinct from the civil power. The role of the ministry in picking and publishing Hoadly's text makes this as much a political project as a religious one. A deluge of pamphlet debate ensued, which was comparable to that instigated by the Longitude Act and was brought to an end only by the advent of the South Sea Crisis.[80] Richard Cocks' contribution epitomised the plays on language that allowed longitude as a project to be linked, and in this case applied, to other projects. The idea of shipwreck allowed him to make his discussion of dissenting religion equivalent to practical navigation. Likewise, his interest in the reward money followed the common theme we have seen above, of suggesting that a comparable reward be offered for solving a range of other projects. The implication is clear: that the solution of the longitude problem is as unlikely as the creation of a united church.

Cocks was, perhaps unsurprisingly, a controversial figure. He was satirised as 'Cockbrain' and 'a whimsical, craz'd man' and was known in the House

79 Cocks, *Perfect discovery*, pp.7–8.

80 For this controversy see Starkie, *Church of England*.

of Commons for his eccentricity, and rambling speeches. His language was difficult for others to follow, yet he emphasised himself as a 'plain man', using his own paths through reason.[81] In 1722, his pamphlet attracted a direct reply from 'Philo-Deus' entitled *Considerations on the Latitude of Sir Richard Cocks's New Opinions* in which he accused Cocks precisely of being a projector and of using the types of projecting language that the Scriblerians feared. He described Cocks' pamphlet as 'this senseless, incoherent scraped Calumny, the Gleanings of Popery and Enthusiasm',[82] equating what he saw as Cocks' poor language with his projecting brain, and commented that 'you never can fail of Applause in … *Change-Alley*'.[83] He made direct comment on the danger of such projecting thoughts being enshrined in print, warning how, 'the Excrements of an enthusiastick Brain seated in the Dross of Nature, sometimes we see delivered with as much Assurance as if inspir'd … without the least Shadow of Reason or Argument, and must be true, because in Print'.[84] This was precisely the relationship of malicious projects to hack writing that the Scriblerians feared. Philo-Deus likewise made play on the language relationships of latitudinarian religious ideas to longitude solutions, commenting on the disparity between Cocks' 'Longitudinarian Principles, and Latitudinarian Practice'.[85]

'A Short Epistle to the Longitudinarians'

In all of these pamphlets and satires, the language of projecting was clearly as important as the content. Language, after all, built the 'bubble' by expressing the exaggerated claims and over-inflated confidence of the projector. In his *Essay upon Projects*, Defoe even drew a direct relationship between the authority of language and the systems of financial credit discussed above, by suggesting that anyone inventing language should be punished in the same way as those found forging currency.[86] In the *Tradesman* he likewise argued for a 'plain style' of speaking and writing as the best way for a merchant to establish his credit, both financially and socially. Poor writing was a constant focus of contemporary satire. In his supposed *Letters from a Persian* in 1735, George, Baron Lyttelton (1709–73), Tory opposition politician, poet, and friend of Pope, equated the search for longitude directly

81 Hayton, 'Sir Richard Cocks', pp.226, 229.

82 Philo-Deus, *Considerations*, pp.5–6.

83 Philo-Deus, *Considerations*, p.49.

84 Philo-Deus, *Considerations*, p.23.

85 Philo-Deus, *Considerations*, p.22.

86 Defoe, *Essay upon Projects*, p.237.

with Grub Street poets. In one letter, his correspondent visited Bedlam and discussed one of its inmates who, 'being of too lively an Imagination for the Dulness of Trade … applied himself to Poetry, and neglecting his other Business, was soon reduced to the State I saw him in: But he assured me he shou'd not be long there; for his lucky Confinement having given him more Leisure for Study, he had quitted Poetry, and taken to the Mathematicks, by the means of which he had found out the Longitude, and expected to obtain a great Reward'.[87] Lyttelton's Bedlam hack reminds us of Samuel Clark's fellow inmate in the Fleet ditch.

The most frequent production of such hack writers was for the newspapers that revolutionised contemporary print. The rage for endless information and such literary projecting was, in turn, equated with the longitude problem. In his 1762 autobiography, the actor George Stayley (1727–79) announced that

> I have made … A Discovery if not equal to Longitude, at least not inferior, in my opinion, to the Philosopher's Stone, or a perpetual Motion; for properly attended to, it will account for all the crowded Falshoods of all our public Papers. – the Word News, is composed of the initial Letters of the four cardinal Winds—*North, East, West, South* … Which may emblematically signify, that News is nothing but Air, which wafts over Lies from every Corner of the Seaman's Compass.[88]

He made a direct relationship between the ephemeral nature of longitude, and its fellow natural philosophical projects the philosopher's stone and perpetual motion, and the ephemeral nature of both the material production and the (in) credible contents of the news. In 1769, *The Batchelor: or Speculations of Jeoffry Wagstaffe* proposed a solution to this '*news madness*' whereby a committee of avid news readers would invent the news. He commented in conclusion that 'This project of mine may, perhaps, be treated by some as a visionary one, like the marine chair, the time-keeper, and other attempts to discover the longitude'.[89] We notice that by this point it was not just generic longitude projects but Harrison's and Irwin's specific proposals that were being used as satirical weapons.

Scriblerian writing endlessly made play with the projecting language of such newspapers and pamphlets, as we have seen. The whole persona of Martin Scriblerus was built around the ability to make him both the author of the group's satirical works, and the supposed author of genuine works that they thought absurd. The language of one aped the other, just as projecting

87 Lyttelton, *Letters from a Persian*, p.14.

88 Stayley, *The life and opinions*, ii, pp.12–13.

89 *The batchelor*, ii, pp.153–4.

pamphlets sought to gain credit through the use of natural philosophical language.[90] Swift summed this up in *A Tale of a Tub* when he interjected the narrative with 'an established custom of our newest authors, by including '*a long digression unsought for,* and *an universal censure unprovoked*'.[91] His satire gives a notably apt description of the linguistic approach used by Jeremy Thacker, in the pamphlet of 1714 that we have already considered in the previous chapter critiquing James Clarke's and William Hobbs' imagery. If we look further at the language of Thacker's pamphlet it becomes clear that it is likely an elaborate literary hoax, written by Arbuthnot for his fellow Scriblerians as a satire on longitude projects. Indeed, Martin Scriblerus had already been made the author of Whiston's longitude scheme, showing the group's interest in such projects.[92] Thacker is heavily sarcastic about a surprisingly large number of other schemes and particularly about their literary style.[93] From the very title page, *The Longitudes Examin'd* suggests the sheer proliferation of schemes, meaning that there is more than one longitude, and Thacker plays with the confidence of other projectors by announcing that the pamphlet describes 'a smart pretty machine of my own which I am (almost) sure will do for the Longitude'.[94]

The pamphlet opens with 'A Short Epistle to the Longitudinarians', satirising the literary style of the usual pamphlets, by commenting that 'without Animadversions upon the Attempts of others, I could not swell this to a Six-penny Book, unless I had embellish'd the Recommendation of my new Device with fine Metaphors, and clever Comparisons'.[95] This bears obvious resemblance to Swift's *Digression.*[96] Thacker goes on to satirise the usual claims made by these pamphlets of disinterest in the reward money, saying, 'If it be ask'd why I wrote the Book at all, I'll frankly answer, *That I wanted Money*; and that if I had thought that the Commissioners would have

90 Ratcliff, 'Art to Cheat the Common-Weale', pp.337–65, in particular p.353.

91 Swift, 'Tale of a Tub' in *Major Works*, p.125.

92 Rogers, 'Longitude forged', republished as 'Satire as Mock-Science' in *Documenting Eighteenth-Century Satire*, pp.45–62. In the republication, Rogers notices more of the allusions mentioned here, but fails to identify Clarke.

93 Another suggestion of Arbuthnot's authorship comes from Newton's correspondence. In November 1714, Christopher Wren sent Newton a cipher for a proposal to find longitude by placing a clock in a vacuum. Wren and Arbuthnot had both been members of the committee to publish Flamsteed's star charts in 1705 and were both Fellows of the Royal Society, so might well have discussed such a solution. Newton, *Correspondence*, vi, Letter 1,120, p.193.

94 Thacker, *Longitudes examin'd*, title.

95 Thacker, *Longitudes examin'd*, p.11.

96 Aked commented in 'A Short Epistle to the Longitudinarians' that this pamphlet could have given Harrison lessons in prolixity, pp.299–300.

been prevail'd upon to have given me some, to carry on Experiments, I had never set Pen to Paper'.[97] His description of himself as 'Well-wisher to the *Twenty Thousand Pounds*'[98] likewise reminds us of the relationship to lottery prizes. The whole pamphlet is then made up of digressions and over-inflated statements, taking pages to get to the invention itself. It uses the common forms of hack writing to emphasise how longitude is an absurd project. We have already seen how Thacker advised Clarke to 'consult about [his invention] with the Inventors of a *Perpetual Motion*'.[99] Whether this contribution is indeed by Arbuthnot or by a genuine, witty Yorkshire projector, it nicely encapsulates the problematic status of longitude as a 'project', the central role of projecting language in this, and how longitude would continue to be linked to a whole spectrum of similar contemporary concerns surrounding other projects in society. It was, of course, hard to engage with the projectors without becoming one of them and calling the credibility of your own language into question, helping us to see why Thacker's pamphlet could be so convincing.[100] Yet this raises a further concern: if projecting could be identified most easily by its author's use of language, how could those attempting to find a solution to longitude, like the Commissioners, know what was really a viable or impossible project? Whose language were they to trust?

97 Thacker, *Longitudes examin'd*, p.iii.
98 Thacker, *Longitudes examin'd*, p.iii.
99 Thacker, *Longitudes examin'd*, pp.6–7.
100 Bogel has discussed the problem of balancing identification and distance in *Difference Satire Makes*, pp.41–83. Lynall has discussed this in relation to Swift in 'Scriblerian Projections', pp.8–9.

4

Madness or Genius?

In 1733–34, the Whig physician George Cheyne (1672–1743) published a seminal work entitled *The English Malady* in which he argued that mental disorder—madness—was a sign of intellectual brilliance. Looking at his contemporaries' achievements in science, literature, and politics, he argued that

> this present Age has made Efforts to go beyond former Times, in all the Arts of *Ingenuity*, *Invention*, *Study*, *Learning*, and all the contemplative and sedentary Professions, [meaning that] the Organs of these Faculties being thereby worn and spoil'd, must affect and deaden the whole *System*.[1]

Too much thinking and success had led, for Cheyne, to a weakening of the English nervous system, making it more prone to disorder. Notably, he limited this to the upper echelons of society, to those who benefited from England's commercial success and had the leisure to do the thinking: 'it seldom … can happen, to any but those of the liveliest and quickest natural Parts, whose Faculties are the brightest and most spiritual, and whose Genius is most keen and penetrating'.[2] For Cheyne, this was a positive sign, the mark of a society that was both civilised and intellectual. The keenest genius might have the most disordered nerves, but these simply required to be 're-tuned'[3] to society

1 Cheyne, *English Malady*, p.53.
2 Cheyne, *English Malady*, p.261.
3 Porter, 'Rage of Party', p.46.

in order to be properly understood. He seemed to see little gap between genius and madness that could not be solved by the right use of language.

Cheyne's argument was pragmatic and political. He was arguing against precisely the kinds of fears about projecting discussed in the last chapter but tying these into wider contemporary concerns about mental disorder. In the early years of the eighteenth century, the Tory government was terrified by the enthusiasm shown by projecting, especially by religious dissenters, which they saw as characteristic of the Whig party and its supporters. We have seen how Tory writers like the Scriblerians presented such projects as threatening established institutions, so that sane members of society needed protecting from this corruption. As a Whig, Cheyne responded by trying to reformulate this 'delusion', making it a marker of social success rather than corruption.[4] Credibility would, by his argument, become rooted in precisely the sorts of nervous mental disorders that made projecting dangerous.

One major way of understanding madness and mental disorder in the eighteenth century was through ideas advanced by John Locke (1632–1704) in 1690 in his *Essay Concerning Human Understanding*. Locke, with whom Cheyne agreed, crucially conceived of madness as knowledge-based rather than physiological, reliant on how the patient connected ideas, rather than any physical malady. Locke had argued that

> idiots … are deprived of reason; whereas madmen … seem to suffer by the other extreme. For they do not appear to me to have lost the faculty of reasoning, but having joined together some ideas very wrongly, they mistake them for truths; and they err as men do that argue from wrong principles.[5]

Madness was, therefore, merely wrong reason, and what counted as 'wrong' clearly depended on how you chose to interpret it. Its opposite relied entirely on 'right' perception, which in turn was clearly reliant on effective communication by the reasoner. We see easily how madness as the result of 'bad' reasoning or 'bad' knowledge could, therefore, be a product of the 'bad' language of projecting that we have considered in the previous chapter. This approach to the difference between madness and reason dogged attitudes to longitude in the period. Not only might madmen be the source of 'bad' or mad schemes for solving longitude, but such schemes could themselves lead to madness by encouraging wrong reasoning. Either you were mad to think of solving longitude, or the process itself could drive you mad.

Many of the authors we have already encountered discussing projecting in the context of Bedlam dwelt on these ideas of madness developing from

4 Porter, 'Rage of Party', pp.41–3.

5 Locke, *Essay Concerning Human Understanding*, pp.93–4.

wrong reason based on wrong perception. Thomas Fitzgerald (c.1695–1752), in his poem *Bedlam*, discussed the reliance of reason on accurate sensory perception, saying,

> the Sense from various Objects brings,
> Through Organs craz'd, the Images of Things;
> Ideas, all extravagant and vain,
> In endless Swarms croud in upon the Brain:
> The cheated Reason True and False confounds,
> And forms her Notions from fantastic Grounds.[6]

He also specifically mentioned longitude within his discussion of Bedlam's inmates:

> Behold a Sage! Immers'd in Thought profound;
> For Science He, for various Skill renown'd …
> All problems has his lively Thought subdu'd,
> Measur'd the Stars, and found the Longitude.[7]

Longitude, it seems, was particularly ripe for encouraging wrong reason. Swift had, likewise, established madness as a product of intense thought in *A Tale of a Tub*, discussing religious 'projecting' as disordering Peter's reason, who 'had kept his brain so long and so violently upon the rack, that at last it *shook* itself and began to *turn around* for a little ease. In short, what with pride, projects and knavery, poor Peter was grown distracted'.[8] A later set of *Miscellanies* claiming to be by Swift, which appeared in 1723, took his ideas even further, linking bad reason and imagination particularly to longitude, saying, 'as free Agents, we have a Right to examine all Things by the Faculties that Heaven has lent us for that Purpose; which are the Reason and the Imagination; so that when one has run it's [*sic*] proper Lengths, we are Morally oblig'd to give a Loose to the others', and 'a Rotation of Intellects … will produce a deal of Speculation upon Circles, and the Squaring of the Comets … the Longitude, the perpetual motion'.[9]

Projectors themselves complained of the risk of their ideas becoming disordered. The map projector William Blundel, whom we met in Chapter 1, complained of such outcomes in his critique of Wilson and Haselden, bemoaning, 'what have I with diligent search, great pains, Labour, Study, and Charges, hazarded my Health, Head, Life, and Estate … all these must

6 Fitzgerald, 'Bedlam', p.3.
7 Fitzgerald, 'Bedlam', pp.8–9.
8 Swift, 'Tale of a Tub' in *Major Works*, p.116.
9 *Supplement to Dean Sw—t's miscellanies*, pp.11–13.

unavoidably create great troubles and afflictions to my mind!'[10] Another critic in the *Weekly Journal* in 1723 commented of Haselden that 'the New Moon … had some influence on his Brains (or what he had in his Head instead of Brains)'.[11] Throughout the eighteenth century, pamphleteers linked longitude specifically to two forms of madness that appear repeatedly in wider social discussions: madness as a product of wrong mathematical reasoning, and as the result of the types of imaginative thinking that could become visionary. We have seen already how these could be an extension of both mathematical and religious projecting.

'STUDY YOURSELVES, YOU POOR NIDDY-NODDIES / LET US ALONE WITH THE HEAVENLY BODIES'[12]

Mathematics was a feature of the longitude problem that made it particularly open to the development of wrong reason and madness. Pope established a special relationship between madness and mathematical reasoning in *The Dunciad* where, in a group of liberal arts confined by Queen Dulness in her cave,

> Mad Mathesis alone was unconfined,
> Too mad for mere material chains to bind,
> Now to pure space lifts her ecstatic stare,
> Now running round the circle, finds it square.[13]

A similar linkage was clearly common in wider hack writing. In the Introduction, we encountered the pirate versions of Hogarth's *Rake's Progress* that were produced by hack engravers from memorising the paintings in his studio. The differences between Hogarth's originals and the memorised copies give us wonderful insight into what references his contemporaries saw in the works. The pirate engraving of Bedlam places a much greater emphasis on mathematics in the lunatic's wall drawing than Hogarth had (Figure 3). Besides the drawings and the word 'longitude' this lunatic has drawn an equation on the wall and written 'Signs Tangents', notably wrongly spelt. He has also written the phrase 'the clock does strike by algebra', which is

10 Blundel, *To all and Singular*, p.27.

11 *Weekly Journal* (15 June 1723), p.2578. This continues a standard association of madness with phases of the moon. The author also compares support for the Mercator projection to belief in transubstantiation.

12 *The Fortune-Hunter*, p.85.

13 Pope, 'The Dunciad' in *Selected Poetry*, p.158.

a line taken directly from Samuel Butler's (1612/13–80) famous narrative poem *Hudibras* first published in 1663, but republished (with engravings by Hogarth) in 1725–26.[14] The opening of this poem described the eponymous knight as pretending to great learning and mathematics, but unable to suit his language to the appropriate company and showing dangerous religious enthusiasm:

> In Mathematics he was greater
> Than Tycho Brahe, or Erra Pater,
> For he, by geometric scale,
> Could take the size of pots of ale;
> Resolve, by sines and tangents straight,
> If bread or butter wanted weight;
> And wisely tell what hour o'the day
> the clock does strike, by Algebra.[15]

These hack engravers' equation of Hogarth's longitude lunatic with the well-known character of Hudibras suggests that they saw both as bound up in these contemporary discussions of mathematical madness and disordered language.

This view on the part of the Scriblerians is particularly clear in Swift's *Gulliver's Travels*. We have already seen how he satirised projecting, the Royal Society, and Exchange Alley in Gulliver's visit to the Academy of Lagado, but we should also note what had inspired these projectors' ideas. Gulliver is told by his host, Munodi, that on going up to the Laputan court, the academicians

> came back with a very little Smattering in Mathematicks, but full of Volatile Spirits acquired in that airy Region … fell into Schemes of putting all Arts, Sciences, Languages, and Mechanicks upon a new Foot … none of these Projects are yet brought to Perfection; and in the mean time, the whole Country lies miserably waste … they are Fifty Times more violently bent upon prosecuting their Schemes, driven equally on by Hope and Despair.[16]

We note how it is mathematics that has inspired the projectors, and how it is represented as a foolish, desperate castle in the air, like so many projects. Gulliver has already visited the Laputan court in his travels, so his readers are well aware of the truth of Munodi's criticisms. The court is obsessed with mathematical learning, but at the expense of their ability to communicate:

---

14 Hogarth's engravings were themselves copied from anonymous images published with an earlier edition of *Hudibras* in 1710. See Hallett, *Hogarth*, pp.16–17.

15 Butler, *Hudibras*, i, p.10.

16 Swift, *Gulliver's Travels*, pp.164–5.

> their Heads were all reclined either to the Right or the Left; one of their Eyes turned inward, and the other directly up to the Zenith … the Minds of these People are so taken up with intense Speculations, that they neither can speak, nor attend to the Discourse of others, without being roused by some external Taction upon the Organs of Speech and Hearing.[17]

The Laputans' thoughts are permanently in the clouds, divorced precisely from the physical sensations that for Locke would lead to reason.

In Hogarth's original print, the lunatic might not use the mathematical solutions for longitude that his pirate copiers remembered, but he does draw on the wall a real contemporary longitude proposal by William Whiston: a character who was widely discussed as mad by contemporaries. He has drawn one of Whiston and Ditton's shells being shot out of a mortar, around the diagram of the world. Ditton had died the year after the passage of the 1714 Longitude Act, but we have already seen how Whiston continued to contribute to the longitude debate for decades, proposing further alternative solutions when the projectiles were universally ridiculed. He was repeatedly satirised as mad or actively accused of madness. His prominent position in the London philosophical community as both a public lecturer and a vocal religious dissenter made him ripe for the character of madman and projector. As the Act was in discussion, Abraham Sharp wrote to John Flamsteed about Whiston and Ditton's proposal, commenting astutely that 'two such noted Mathematicians should not amuse the nation with such mistakes and fallacys … since that would be utterly to ruin their Credit which I presume they in a great measure maintain themselves by'.[18] He shows nicely how ideas of credit played into discourses of madness as much as projecting, both relying on careful management of language to establish claims to knowledge and expertise.

Satires on Whiston appeared almost immediately in 1714, and their sheer proliferation helped to establish longitude as a mad project.[19] *Will-With-a-Wisp; or, The Grand Ignis Fatuus of London* specifically satirised his religious beliefs together with his longitude proposal. The writer said of the projectile scheme that

> It looks as yet, only like a *Trick* to get *Money*, upon the wisest *Distrust* of his uncertain Pretensions. A *Sum* of ten or twenty thousand *Pounds* would do fine Things, fill his *Coffers*, or his *primitive Library* with *Arian* books, and peradventure *square the Circle* and all, into the Bargain … I must tell the best

17 Swift, *Gulliver's Travels*, p.146.

18 Flamsteed, *Correspondence*, Letter 1,356, pp.695–7, quotation p.696.

19 Lynall, 'Scriblerian Projections', p.8.

> *Longitudinarian* of 'em all, by the By, that the farthest Way *about*, is very often the nearest Way *Home*, in Point of *Safety*, both by *Sea* and *Land*.[20]

Again, we see the equation of longitude with other types of projecting and with ideas of shipwreck, as well as the emphasis on the reward money. He went on to question Whiston's sanity, saying, 'I do not know for certain how well our *Will-With-a-Wisp* understands *Longitude* after all; and yet I verily believe, to the utmost Perfection, in the wonted Conceit of his own wavering Mind'.[21] In 1721, the controversial theologian Thomas Woolston (1668–1733), who had himself been confined for 'wild' religious ideas, specifically discussed both Whiston's religious and longitude attempts as projecting, saying, 'I believe him to be an excellent Mathematician who adorn'd the Professor's Chair at *Cambridge*; but at the writings of the Primitive Church he has no Talent. He has indeed a plodding and projecting Head … tho' he has not as yet found out the *Longitude*, nor ever will the *Philosopher's Stone*'.[22] We have already met the projector Shirtless in William Hunt's play of 1737, seeking a dowry for his daughters by discovering longitude. On hearing of his scheme, the honest investor comments that 'part of my two thousand Pounds is to be blown up on *Hampstead Heath*, I suppose', clearly referencing Whiston's ongoing projectile experiments there.[23]

Whiston fuelled this satire by constantly involving himself in more longitude schemes than any other single contributor. After his 1714 proposal with Ditton, he published proposals for finding longitude using magnetic variation in 1719 and 1721, as we have seen, and using solar eclipses in 1724.[24] In 1734, he proposed to the Royal Society a new type of reflecting telescope that could be used to observe the eclipses of Jupiter's satellites.[25] He also sent a copy to the Board of Longitude via the Admiralty, which is, perhaps tellingly, labelled in the hand of the Lord High Admiral Sir Charles Wager (1666–1743) 'Mr Whiston's Project for finding the Longitude'.[26] He went on to petition the Admiralty about the invention again in 1737, and to publish

20 N. N., *Will-With-a-Wisp*, pp.70–1.

21 N. N., *Will-With-a-Wisp*, p.72.

22 Woolston, *Second letter*, p.45. For Woolston's history see William H. Trapnell in *ODNB*.

23 Hunt, *The projectors*, p.46. Whiston's experiments on Hampstead Heath are described, including a map, in *To the Honourable the Commissioners* and *The Longitude discovered by the eclipses* (1738).

24 Whiston, *The longitude and latitude found* (1719), *The longitude and latitude found* (1721), and *The calculation of solar eclipses*.

25 'Description of Mr Whiston's Reflecting Telescope', Royal Society Archives, RBO 19/1.

26 'Mr Whiston's Project', Caird Library, MSS/79/130.2.

an extended discussion of his various means of finding longitude in 1738 and an expanded edition in 1741, both entitled *The Longitude Discovered by the Eclipses, Occultations and Conjunctions of Jupiter's Planets*.[27] The manuscript description read to the Royal Society in 1734 suggests that Whiston was not impervious to the satirical accusations of madness that were levelled at his persistent interest in longitude, nor to concerns over the impossibility of so many schemes. He discussed how he had shown the telescope to the Society in 1730, but 'I did not then carry this discovery any farther; because I was determined not to publish any thing more about the Longitude, till I was fully satisfied that the method was not only right, but also would be practicable at Sea'.[28] We have seen that Whiston would eventually be awarded five hundred pounds by the Commissioners of Longitude to map the British Coast using magnetic variation, yet even here he did not escape satire. In 1750, the prolific mathematical writer Robert Heath (of whom more later) published *A Natural and Historical Account of the Islands of Scilly* in which he commented, 'As to Mr *Whiston*'s Survey of the *English* Channel (which pretends to correct Dr. *Halley*'s) it is inaccurate in many Places about *Cornwall*, and not to be depended on ... equally certain with the Event of that learned Gentleman's late *Prophecy* of the *Period* to this Scene of all Things, and his Discoveries concerning the *Longitude*'.[29] He references Whiston's seemingly dogged belief in the prophetic nature of comets, for which the Scriblerians also particularly mocked him.[30]

Whiston was, indeed, a continued figure of fun for the Scriblerians. Swift and Arbuthnot almost crowed with joy in letters to each other over their inclusion of his scheme in Martin's *Memoirs*.[31] Arbuthnot wrote to Swift in 1714 that Whiston had stolen one of his ideas for Scriblerus by publishing his proposal,[32] and Swift answered that 'It was a malicious Satyr of yours upon Whiston, that what you intended as a Ridicule, should be any way struck upon him for a Reality'.[33] The Scriblerians are also credited with writing a scatological ode against Whiston and Ditton's scheme that gained wide public appeal and continued use. The simple point of the ode may be summed up by the first two verses:

27 Whiston, *The Longitude discovered by the eclipses* (1738) and the same title from 1741.
28 'Description of Mr Whiston's Reflecting Telescope', pp.1–2.
29 Heath, *A natural and historical account*, p.451.
30 Rousseau, '"Wicked Whiston"', pp.17–44.
31 Kirby-Miller, *Memoirs*, pp.166–7.
32 This is one of many elements that persuades Rogers that Arbuthnot was Thacker.
33 Arbuthnot, *Correspondence*, Letter 64, Jonathan Swift to Arbuthnot, 25 July 1714, p.195.

RECITATIVO
The Longitude mist on
By wicked Will. Whiston.
And not better hit on
By good Master Ditton.

RITORNELLO.
So Ditton and Whiston
May both be bep-st on;
And Whiston and Ditton
May both be besh-t on.[34]

The Rawlinson Collection of ephemera at the Bodleian Library in Oxford includes a hand-written version of the ode on a scrap of paper.[35] This suggests an oral tradition in which the wording of the ode changed subtly. Here the description of Whiston continued to be 'wicked', while Ditton became 'honest Num', suggesting how Whiston's reputation continued to suffer from his interest in longitude. John Gay went on to satirise Whiston's mixing of natural philosophical and mathematical work with Christian theology in *A True and Faithful Narrative of what pass'd in London*. In this, he used the idea of Whiston prophesying the imminent destruction of the world by a comet in one of his coffee-house lectures to critique the superficial adoption of religious devotion by society in such circumstances.[36] There are clear parallels here with the projects of Martin Scriblerus more generally, as well as with Arbuthnot's 1722 satire *Annus Mirabilis*, in which he imagined the impact of a comet causing the sex of all men and women in society to swap.[37] Notably, Arbuthnot's signature as 'Well-Wisher to the Mathematics' uses the same formula as Thacker signing himself 'Well-Wisher to the Twenty-Thousand Pounds', as well as emphasising the mathematical nature of his satirical prophesy. While there is, of course, larger social satire in play here, we can see the way that Whiston had become a figure used to symbolise wider public madness in the face of mathematics.

The Scriblerians clearly relished the chance to exploit such a figure of eccentricity as Whiston in their satires. Yet it is unsurprising to some extent that Swift, for one, was so mistrustful of longitude as a mathematical problem, given that he had personal experience of its vicissitudes. An Irish friend, Joseph Beaumont (fl.1700–d.1727), went mad from just such

34 'Ode, for Musick, on the Longitude' in *Miscellanies*, p.172.

35 'A Short Ode', Bodleian Library, MSS Rawl Poet 152, f.97. My thanks to Leanna McLaughlin for sending me this information.

36 Gay, 'A True and faithful Narrative', ii, pp.464–73.

37 Arbuthnot, *Annus Mirabilis*.

mathematical inquiries, with Swift attempting to commit him to Bedlam while Beaumont was in London in 1722, before the poor man ultimately committed suicide.[38] As early as April 1715, a mutual friend, Knightley Chetwode (1650–1720), wrote to Swift, 'For God sake do somewhat to comfort Joe, and send him back contented ... help him in his longitude, do anything to keep him alive'.[39] This is presumably the context to Swift's reply to a request from a John Wheldon (fl.1729) in 1729 that Swift would use his influence with some of the Commissioners of Longitude to get Wheldon their attention.[40] Swift replied that 'I understand not Mathematicks, but have been formerly troubled too much with Projectors of the Longitude to my great Mortification and some Charges by encouraging them ... One of my Projectors cut his Throat, and the other was found an Imposter ... [I] am confident you would deceive others, or are deceived yourself'.[41] He summed up perfectly the fear that longitude proposals were either projects, or the product of a brain deceived by too intense mathematical thought.

Most fascinatingly, by 1758, ideas of mathematics and longitude as leading to madness had entered medical discourse directly. In that year, William Battie (1703–76), the innovative new physician at St Luke's Hospital for Lunatics, published his *Treatise on Madness*, which was largely a critique of the old-fashioned and Draconian treatments used at Bedlam. In this, he discussed 'the chimærical dreams of infirm and shattered Philosophers, who after having spent many days and nights without closing their eyes in unwearied endeavours ... to discover the Longitude or grand Secret, have at last fallen asleep, and who by excessive attention of body have strained every animal fibre, and may without a metaphor be said to have cracked their brains'.[42] He seems to respond directly to these discussions of longitude as linked to madness in his conception of the inmates and conditions at Bedlam.

'FOR YOU MUST DO THIS UPON YOUR KNEES'

MANY, THEREFORE, CLEARLY SAW in Hogarth's Bedlam image the same mathematical madness that his pirate copyists remembered. However, it is noteworthy that the poem that accompanied the piracies

38 Lynall, 'Scriblerian Projections', p.9.

39 Swift, *Correspondence*, Knightley Chetwode to Swift, 25 April 1715, p.165.

40 Swift, *Correspondence*, John Wheldon to Swift, *Dublin Journal*, 31 May–3 June 1729, [September 1727], pp.239–40.

41 Swift, *Correspondence*, Swift to John Wheldon, p.240.

42 Battie, *A Treatise*, pp.56–7.

(Figure 3), emphasised a different aspect of our lunatic's madness: the imaginative, visionary kind. The poem commented of the inmates:

> What strange Chimeras fill the Brain,
> When Reason does no longer Reign!
> As Frenzy flows from different Springs,
> Men mimick Poets, Popes or Kings
> With superstitious fears sit quaking
> Or Combat Devils of their own making.[43]

Again, we see the concern with reason, but now superstition and devils are also part of the mental landscape, bringing in a religious dimension. We have seen the importance of religious enthusiasm in discussions of Richard Cocks' project for 'the longitude of the church' in Chapter 3 and in the interplay of Whiston's dissenting and mathematical views. Our longitude lunatic is accompanied in Bedlam by the mad Pope, whose make-shift crosier strikingly resembles a cross-staff, linking him both to navigation and to the types of mathematical instruments that were iconographically associated with melancholic madness.[44] Both are, further, accompanied by the religious enthusiast, who himself gazes enthralled at his cross bathed in a beam of light (Figure 1). This religious inmate of Bedlam suggests the more visionary, inspired form of madness, which also appeared as a facet of the longitude problem.

A number of contributors to the longitude debate argued that the search for longitude was itself an act of religious devotion, fulfilling God's purpose, but only at the appointed time, in the manner of one of Whiston's prophetic comets. In 1732, Richard Locke (fl.1730–52), another prolific contributor of longitude proposals, apologised for any mistakes in his publication, explaining that 'God, to discover more his own Perfections, and to give us greater Ideas of his infinite Wisdom and Majesty, has concealed even those visible Things from the deepest Search of the most profound Philosopher: All that they can do, is to erect Schemes in their own Imagination'.[45] We see immediately the dangers of mental disorder that such imaginative schemes might create. John Hanna (fl.1725–36) published a proposal in 1725, in Dublin, which highlights how some proposals linked religious and natural philosophical discoveries directly. His pamphlet title shows this instantaneously: *An Astronomical Creed … with*

43 The phrase 'Chimeras fill the Brain' interestingly echoes Charles Leadbetter's description of a 1730 proposal by John Bates as 'by Chimaeras in his Brain' in a list of longitude proposals in *Uranoscopia*, p.310.

44 Such attributes of Melancholy have most famously been discussed by Panofsky in *Saturn and Melancholy*, pp.317–73.

45 Locke, *The Longitude*, p.24.

Natural and Moral Reflections ... to which is Subjoined, the New Method of Discovering the Longitude, by the Appulse of the Moon to the Fixed Stars. He went on to explain how he had built this on Newton's lunar theory and Flamsteed's work on parallax.[46] His introduction emphasised the spiritual nourishment that he saw in such natural philosophical enquiry, arguing that 'to be free from the Intanglements of prejudice and errour, and to have new light shining into the mind is the excellency and pleasure of a Rational Soul'.[47] He proceeded to critique other longitude proposals, including of course Whiston and Ditton, advocating his own astrological solution. Interestingly, he commented directly on language use, criticising the over-inflated language of projecting pamphlets and newspapers: 'methinks that ... they that publish false News, ought to be Punished as abusers of that useful method of Publick Intelligence, so ought they, who are guilty of such impudent Impositions on the weakness of others'.[48] By contrast, with his astrological solution, 'it's necessary to speak beyond what is true to bring the Idea of truth into the mind, which is or should be the design of all Speech'.[49] Again, a problematic proposal when it comes to reasonable language.

A better-known longitude visionary was the Swiss natural philosopher and friend of Newton, Nicolas Fatio de Duillier (1664–1753), who lived in England in the 1690s–1720s. As a member of the Royal Society, he was known for his heretical, enthusiastic, and visionary interests and was seen largely as a charlatan. Robert Hooke named him 'the Perpet[ual] Motion man'.[50] Much like Whiston, he combined a range of natural philosophical interests that related to longitude schemes, working particularly on the use of jewels in watch-work, and on a magnetic cause for gravity, the latter of which saw him collaborate with Newton. He also became involved with religious prophecy and ideas of natural philosophy as an expression of divine providence. By 1707, he was known to be part of an immigrant missionary group called the 'French Prophets'. He carried out missionary work in Europe and entwined his natural philosophical ideas with the Prophets' teachings. He corresponded on these matters with Whiston himself, arguing for the importance of his own ideas in influencing Newton's views on longitude.[51]

46 Hanna, *Astronomical Creed.*

47 Hanna, *Astronomical Creed*, p.3.

48 Hanna, *Astronomical Creed*, p.43.

49 Hanna, *Astronomical Creed*, p.48.

50 Mandelbrote, 'The Heterodox Career of Fatio de Duillier', pp.269–77, Hooke quoted on p.277.

51 Faccio, *Letter*, Beinecke Rare Book and Manuscript Library, Gen MSS FILE 85. On Fatio's interactions with the British natural philosophical community see Mandelbrote.

Fatio saw his mechanistic ideas of gravity as part of enthusiastic religion, proof of God's constant intervention in the world, and a means to announce a new union of earth and heaven.[52] Fatio's interest in the longitude problem is evident from the rare surviving abstract of the 1714 Act that is pasted into his copy of the first edition of Newton's *Principia*.[53] He has annotated the abstract, underlining the key elements of the Act, calculating the error in miles allowed for each reward, and commenting that, by 1738, all nine of the Commissioners named in the Act were dead. Despite his varied interests, Fatio only once published a solution to the longitude, in 1728. In his *Navigation Improv'd*, he explained that

> Among the many and useful things, which Providence and my long and too unfortunate and ruinous Study of the Mathematicks has enabled me to find ... I shall chuse to publish now chiefly one ... By which I shall invite the Generosity and Justice of this flourishing Nation, in pursuance of the Acts of Parliament.[54]

We note his querying, like Robert Browne, that the Commissioners seemed unwilling to judge proposals and award money, along with his equation of the workings of providence with his 'long and too unfortunate and ruinous Study of the Mathematicks'.

Such study could indeed end in the madhouse. It was in the very St Luke's Hospital where William Battie presided that the visionary poet Christopher 'Kit' Smart (1722–71) spent six years between 1757 and 1763. There, he composed his famous poem *Jubilate Agno*, legendarily writing with chalk or a key directly onto the walls of his cell.[55] In the poem, Smart discussed natural philosophy as a path to spiritual redemption. He showed a supremely Augustan view of man as a fallen sinner, whose misapprehensions and wrong visions of the world were caused by his break with God. It was, therefore, in improved observation that his salvation lay: in how nature was perceived, and how man's acts of vision were assessed.[56] Within the poem, Smart discussed the longitude problem as an example of how natural philosophy needed to incorporate religious belief to create a positive interpretation of nature. He referenced the different methods proposed so far, including both quadrants

52 The enthusiastic side of Fatio's career has been discussed by Heyd in *Be sober and reasonable*, pp.242–61.

53 *An Abstract*, Bodleian Libraries, Arch. A d.37. My thanks to Sarah Wheale and Susannah Gibson for help with this abstract.

54 Fatio de Duillier, *Navigation improv'd*, p.3.

55 This legend is mentioned by Kernan in *Printing Technology*, pp.82–4.

56 Guest, *Form of Sound Words*, pp.196–240.

and magnetic variation,[57] but it was observation of the stars that would bring a solution:

> *For I bless GOD in the discovery of the LONGITUDE direct by means of GLADWICK …*
> Let Jonas rejoice with the Sea-Devil, who hath a good name from his maker.
> For GLADWICK is a substance growing on hills in the East, candied by the sun, and of diverse colours.[58]

It is likely that 'gladwick' was mica, the 'isinglass' described by Newton in his *Opticks* as having unusual properties of refraction.[59] Smart interpreted this as meaning that the crystal had a magnetic attraction on light, which would enable it to find true East, therefore longitude, and therefore Eden:

> *For the LONGITUDE may be discovered by attending the motions of the Sun …*
> *For you must consider the Sun as dodging, which he does to parry observation.*
> *For he must be taken with an Astrolabe, and consider'd respecting the point he left.*
> *For you must do this upon your knees and that will secure your point.*[60]

Thus, for Smart, the solution to longitude was literally visionary, requiring the use of 'gladwick' as a new sort of lens for observation at sea. He combined practical and quasi-Newtonian natural philosophy with a belief in the need for spiritual regeneration. 'For you must do this upon your knees' advised the sailor as much on the need for a steady observation, as the natural philosopher on the need for spiritual belief. Only by combining both would he 'secure his point' of longitude.

'True Genius walks along a line'

We have already seen how Samuel Johnson incorporated the problematic relationship of longitude and projecting into his *Dictionary* through his choice of quotations, but he also showed a more complex relationship with learning. In defining '*literati*, the learned' he used

57 Smart, *Poetical Works*, Fragment B, pp.39–40, lines 168–9.
58 Smart, *Poetical Works*, Fragment B, pp.43–5, lines 190–9.
59 Guest, *Form of Sound Words*, pp.235–7.
60 Smart, *Poetical Works*, Fragment B, p.65, lines 349–52.

a quotation from Richard Steele's (1672–1729) *Spectator* that 'I shall consult some literati on the project sent me for the discovery of the longitude'.[61] This related projecting to 'the learned', and the learned we have seen were a broad and disparate group. From Arbuthnot to Smart, via Flamsteed and Whiston, they were all interested in the longitude problem, all believed in particular types of solution, and all viewed alternatives as varyingly ridiculous, mad, or dangerous. The longitude problem focused the mind on a complex, particular project to an extent that might lead to the reason becoming disordered, but it would nonetheless take a new association of ideas in order for any solution, in fact, to be found. Whether or not a solution was a project or a genuine proposal, whether its author was disordered or not, mad or a genius, would not necessarily be visible given that the solution to the problem was unknown. It would take 'vision' to bring any solution to perfection, let alone to inscribe it on paper, or bring it to practicality and usefulness. Oliver Goldsmith (1728–74) put this conundrum perfectly in his anonymous *The Bee* in 1759:

> An author, who would be sublime, often runs his thought into burlesque; yet I can readily pardon his mistaking ten times for once succeeding. True Genius walks along a line, and, perhaps, our greatest pleasure is in seeing it so often near falling, without being ever actually down. Every science has hitherto undiscovered mysteries, after which men should travel undiscouraged by the failure of former adventurers … We shall, perhaps, never be able to discover the longitude, yet, perhaps, we may arrive at new truths in the investigation.[62]

Thus, he cited longitude as precisely the sort of line on which genius balanced.

Michel Foucault (1926–84) beautifully described the complex status of the madman or genius's knowledge as 'an iridescent bubble'.[63] The boundary between rightly connected ideas, wrongly connected ideas, and merely newly connected ideas was so very fine, so ephemeral and fragile, that it was as hard to hold onto as the skin of a bubble, as ephemeral as the projects we have discussed above. This problematic border was also recognised by many of our writers on madness and madhouses. Ned Ward's anecdotal tales of the city, *The London Spy*, included the particularly piquant assertion by a Bedlamite that only the mad really saw truth:

> we Madmen have … much Priviledge of speaking our Minds, within these Walls, … Truth is Persecuted every where Abroad, and flies hither for sanctuary, where she sits as safe as a Knave in a Church.[64]

---

61 Johnson, *Dictionary*, ii, 4D2v.

62 *The Bee*, pp.120–1.

63 Foucault, *Madness and Civilisation*, p.22.

64 *London Spy*, p.65.

The main theme of Swift's *A Tale of a Tub* was, likewise, essentially the point at which philosophy became madness, and even Cheyne's fashionable mental disorders came only to 'those of the liveliest and quickest natural Parts, whose Faculties are the brightest and most spiritual, and whose Genius is most keen and penetrating'.[65] Earlier in Tom Rakewell's story, in Newgate prison, his cellmate is shown building a pair of wings in the corner of the room. These are, clearly, another impossible project—the attempt to fly—but also reference the wings of inspired genius that had accompanied figures of melancholic madness since at least Albrecht Dürer's (1471–1528) *Melancholia* engraving in 1514, and which take us back to Cibber's figures of melancholic and raving madness above the gates of Bedlam.[66]

Genius was difficult to establish, therefore, because of its similarities with madness, and how the boundary between the two seemed to shift in different companies.[67] What made a philosopher a genius rather than a madman? For contemporaries, it seemed to depend on two key concerns. The first was the relationship between the senses, internal reason, and how this was then expressed to others. The important thing was to 'tune' your language and ideas such that they chimed with the expectations of your audience. Again, Swift gave the clearest explanation of this. In *A Tale of a Tub*, he explained,

> there is a peculiar *string* in the harmony of human understanding which, in several individuals, is exactly of the same tuning ... if you chance to jar the string among those who are either above or below your own height, instead of subscribing to your doctrine they will tie you fast, call you mad, and feed you with bread and water. It is therefore a point of the nicest conduct to distinguish and adapt this noble talent, with respect to the differences of persons and times ... [for] it is a fatal miscarriage so ill to order affairs as to pass for a *fool* in one company, when in another you might be treated as a *philosopher*.[68]

He took this further in *Some Thoughts on Free Thinking*, where he laid out the importance of choosing which ideas to express:

> the difference betwixt a mad man and one by his wits ...[is] that the former spoke out whatever came into his mind, and just in the confused manner as his imagination presented the ideas: The latter only expressed such thoughts as

65 Cheyne, *English Malady*, p.261.

66 My thanks to Alexander Marr for help with these ideas. See Panofsky, *Saturn and Melancholy*. Wings also link to James Amelang's work on artisan autobiography, *Flight of Icarus*.

67 Fara has discussed such changing ideas of genius in specific attitudes to Newton in *Newton*.

68 Swift, 'Tale of a Tub' in *Major Works*, p.142.

> his judgment directed him to chuse … in the great multiplicity of ideas which one's mind is apt to form, there is nothing more difficult than to select those which are most proper.[69]

Likewise, Cheyne's discussion of English 'social' madness essentially required disordered nerves to be 're-tuned'[70] to society. Given these concerns, it is amusing to find that Swift would himself 'tune' his conversation to an audience. His friend the clergyman Patrick Delany (1686–1768) wrote in 1754 how 'he could so ill bear to be considered as a cipher in any *scientific* society, that he applied himself even to mathematics'.[71] Perhaps even mathematics was necessary not 'to pass for a *fool*' in some circumstances.

Longitude pamphleteers generally approached this conundrum from two angles, either through an apology for their style, or through asserting that their solutions stemmed from unaided genius, divorced from traditional education, of which a disordered style was therefore proof.[72] As early as the 1690s, the mariner Edward Harrison (fl.1696), who accompanied Halley on his first *Paramore* voyage, commented on his language in comparison to that of scholars: 'I understand not how to Express my Sentences, in … *Sophistical Rhetoric*; if you meet with a plain Method, downright Reasoning, and matter of Purpose, it is what I aim at'; and he justified the status of his knowledge, saying, 'if I know more than others, it is by *Divine Authority*, by *Industry*, and *Experience*, by an *Inborn* Idea, and Instinct in *Nature*'.[73] We might think back to William Derham's concern over addressing both artisans and gentlemen in the *Artificial Clock-Maker*. Two particular longitude candidates had their youth or lack of conventional education emphasised as a mark of their unusual abilities and genius. In 1732, James Logan (1674–1751), the chief justice of Pennsylvania, wrote to Edmond Halley about a young local glazier called Thomas Godfrey (1704–49)

> who had no other Education than to learn to read and write with a little common Arithmetic, having in his Apprenticeship … accidentally met with a Mathematical Book, took such a fancy to the Study, that by the natural Strength of his Genius without any Instructor, he soon made himself master of that, and of every other of the Kind he could borrow … and finding there was more to be had in Latin Books, under all imaginable Discouragements applied

69 Swift, 'Some Thoughts on Free Thinking', pp.14–15.
70 Porter, 'Rage of Party', p.46.
71 Delany, *Observations*, p.101.
72 Amelang notes that apologies for style were often used in such works, *Flight of Icarus*, p.159.
73 Harrison, *Idea Longitudinis*, pp.8–9.

> himself to the Study of that Language … after which, the first time I ever to my knowledge saw or heard of him, he came to borrow Sr. I: N's Principia.[74]

Godfrey told Logan that he had been working on a reflecting instrument for taking observations at sea and showed him a model of a sea quadrant. Logan commented to Halley that he had never before seen a comparable instrument and recommended Logan as a candidate for the longitude reward.[75] John Philip Baratier (1721–40), by contrast, was an exemplary youthful German prodigy who could speak several languages, both ancient and modern, disputed with learned philosophers while still a child, and published eleven works in his short life.[76] At age thirteen, Baratier was reported to have discovered globes and been completely distracted by them:

> He made an Astrolabe, and composed astronomical Tables … He invented new Methods of Calculation, or such as were new to him, since he found them not in the Authors he consulted … At this time he devised his Project for the Discovery of the Longitude, which he communicated to the Royal Society at *London* … in *January* 1735, that is, about three Months after the Globes fell into his Hands.[77]

This was used as a key example of Baratier's prodigious genius, more so than his work in languages and literature.

Samuel Johnson himself helped another longitude projector to tune his voice to his surroundings: the ageing and infirm Zachariah Williams (1673–1755). In the 1750s, Johnson first advertised Williams' scheme for finding longitude by magnetic variation in a series of letters to the *Gentleman's Magazine*,[78] and then co-wrote *An Account of an Attempt to Discover the Longitude at Sea*.[79] This reads much more like a literary than a scientific text and has a markedly Johnsonian melancholic ring to it. The pamphlet recounted the history of Williams' work on the longitude at length, emphasising his hard work for no reward, as we have seen all the best projecting pamphlets did. It likewise discussed the unsuccessful attempt to create a small group of investors to

74 'James Logan', Royal Society Archives, MS EL/L6/59, p.1.

75 Godfrey's quadrant pre-dates Hadley's discussed in Chapter 2, and therefore seems to present a genuine example of the instrument being developed by two independent thinkers at the same time. On the complicated relationship of these inventions see Bennett, 'Catadioptrics and Commerce' and Gerrard, *Astronomical Minds*.

76 Formey, 'Life of John Philip Baratier', pp.245–87.

77 Formey, 'Life of John Philip Baratier', pp.255–6.

78 See 'Original Letters of Zachary Williams' in the *Gentleman's Magazine* 57.3 (September 1787), pp.757–9 and 57.6 (December 1787), pp.1,041–2.

79 Albert Kuhn discusses this relationship in 'Dr Johnson, Zachariah Williams', pp.40–52.

support Williams in his experiments.[80] One episode narrated how Williams had been encouraged to approach the Admiralty in 1729 and, supported by Lord Torrington (1663–1733), had met and shown his scheme to Newton who had 'believed my System visionary' but that 'I am able to produce, in my Favour, the Testimony of Time, the inflexible enemy of false Hypotheses; the only Testimony which it becomes human Understanding to oppose to the authority of *Newton*'.[81] Johnson set Williams' devotion to his work and slow, steady discoveries divorced from the natural philosophical community, against the known eccentricity of Newton the genius. He ended with the plaintive comment that Williams found himself 'a kind of Stranger in a new World, where Curiosity is now diverted to other Objects, and where having no Means of ingratiating my Labours; I stand the single Votary of an obsolete Science, the Scoff of puny Pupils of puny Philosophers'.[82] He had, essentially, lost the ability to communicate his ideas at the right pitch for his audiences, and thus become a fool where he had tried to be a philosopher.

The second concern marking the boundary between madness and genius was the question of originality or replication. In his 1759 *Conjectures on Original Composition*, Edward Young (1683–1765) argued forcefully that genius was not replicable. He established originality as the hallmark of genius and as natural knowledge creation: 'the mind of a man of genius is a fertile and pleasant field ... it enjoys perpetual spring. Of that spring, *Originals* are the fairest flowers'.[83] By contrast, imitation or replicated knowledge was mechanical and the product of learning rather than genius: '*Imitators* only give us a sort of duplicate of what we had ... increasing the mere drug of books, while all that makes them valuable, *knowledge* and *genius* are at a stand'.[84] His discussion referred solely to literary works, but the language that he used to discuss genius as natural or mechanical obviously relates to instrumental or mathematical solutions to the longitude problem, as well as to the 1769 discussions about authorship rights in machines and texts outlined in Chapter 2. Likewise, we have seen how longitude projecting and language were often considered from a linguistic and literary perspective. The problem for writers like the Scriblerians was that new projects were usually associated with new projecting language and printed forms. So, the question of communication became not only tuning one's language to the appropriate audience, but also the question of original language and

80 Williams, *Account of an Attempt*, p.8.
81 Williams, *Account of an Attempt*, p.14.
82 Williams, *Account of an Attempt*, p.16.
83 Young, *Conjectures*, p.9.
84 Young, *Conjectures*, p.10.

its relationship to original ideas.[85] Such questions of innovation are clearly eminently pertinent to the period of the longitude debate. Any potential longitude solution would require 'an original', in terms equally of thinker, machine, and communication, but it would also need unquestionably to be replicable in order to become 'practicable and useful at sea'. If it remained unrepeatable, it could not gain the reward.

Such concerns were clear to those who attempted to solve, or satirised, the problem in their pamphlets. In 1723, in *Considerations on the Unhappy Accidents*, a work on natural philosophical projects, the author commented of a potential mechanical solution to longitude that 'experience has taught, that it is impossible for any Artificer to make Two Engines exactly alike; therefore it is requisite to seek for an *Authomatom*, whose Inequality we are able to find out, that we may, by perceiving the Difference, come to a knowledge of the Exactness'.[86] In this case, replication was necessary in order to reduce an invention from the mind of a genius to practical application. In 1764, in *The Triumvirate*, his imitation of *Tristram Shandy* (of which more in Chapter 6), the Irish writer Richard Griffith (fl.1764–d.1788) bemoaned, by contrast, the lack of originality in contemporary literature and compared the search for longitude to this favourably, saying,

> let us but look back into the annals of literature, for half a century past, and we shall hardly meet with any thing new, except the type, ink, and paper. Nothing but former writings dressed up again, and *re-published*. No work of genius, no useful discovery; except ... one or two attempts toward discovering the *Longitude*, at sea.[87]

So, in fact, the slow, steady attempt to solve the longitude problem might lift endlessly replicated literature out of the depths of printed projecting and enliven it with a little genius. The moralist author John Browne (b.1695/96–fl.1731) had summed up such problems in 1730, in a pamphlet dealing largely with religious concerns. He commented that 'Warm Men can *square the Circle*, and settle the *Longitude*, and *demonstrate* what they please with all imaginable Readiness. But cooler Men will be content to stop within the Bounds of clear and distinct Perception, and will not pretend to *Demonstrations* beyond their Ideas'.[88] He emphasised how a solution to the longitude would not stay within acceptable ideas and, therefore, essentially would not be solved by reasonable men.

85 Bogel, *Difference Satire Makes*, pp.215–47.
86 Gentleman, *Considerations on the unhappy accidents*, pp.31–2.
87 Griffith, *Triumvirate*, i, p.262.
88 Browne, *Defence of the Bishop of London*, p.48.

'either he must be out of his Senses, or I must be so!'[89]

The question of what type of 'right' reasoning could solve the longitude problem proved most problematic when potential solutions came to be considered by the Board of Longitude. In order to merit the coveted 'great reward' a solution had not only unquestionably to establish a new means of finding longitude accurately at sea, but also to be replicable and to be understood across a spectrum from the Astronomer Royal to the navigator of a ship. The problems of language and communication that affected the borderline between proposals and 'projects', madness and genius, are therefore unsurprisingly very visible in the Commissioners' interaction with John Harrison, and in how he was represented and discussed in contemporary literature. The mathematical teacher and naval officer John Hamilton Moore (1738–1807) summed up the situation perfectly in 1772 in his *Practical Navigator*, putting Harrison's achievements and struggles in the context of precisely the discourses we have been discussing. He explained that

> The Difficulty in ascertaining the Long[itude] At Sea has inclined many to attempt it … much the greater Number … were visionary whimsical Men, who, being ignorant of the Difficulties attending such a Discovery, and lured by the Hopes of the appointed Reward, have shut themselves up in a Closet to invent what every Man of Sense must laugh at. / But that great Artist Mr. Harrison has invented a Machine to come nearest to that wanted … but whether it's [*sic*] Composition will admit of being rendered of general Use, I cannot say.[90]

Harrison has commonly been presented by authors from his own time to ours as a natural or 'lone genius'.[91] His exceptional skills in clock-making, developed from a background in carpentry and seemingly isolated from the clock trade, play perfectly into just such tales of uneducated genius as we have already seen in other parts of the longitude story. Undoubtedly, this narrative comes largely from the account of Harrison's life published after his death, which recounted a story that mixes the uneducated genius with the child prodigy. These described how 'he was, from his early childhood, attached to any machinery moving by wheels, as appeared while he lay sick of the small-pox, about the 6th year of his age; when he had a watch placed open upon his pillow, to amuse himself by contemplating on the movement'. From

89 Harrison, *Description*, p.7.

90 Hamilton Moore, *Practical Navigator*, p.189.

91 Sobel, *Longitude*.

this example of early interest, the account went on to emphasise 'the vigor of his natural abilities ... though his opportunities of acquiring knowledge were few, he eagerly improved every incident from which he might collect information; frequently employing all, or great part of his nights, in writing, or drawing ... His native genius asserted itself superior to these solitary disadvantages'.[92] In the context of our discussions in Chapter 2 we note here the emphasis on Harrison's commitment to learning and practising, writing, and drawing. We also learnt in that chapter that he copied a set of lectures borrowed from the local clergyman.[93] Yet it is worth remembering that Harrison was in his mid-thirties by this time, and thus the child genius narrative that echoes Godfrey and Baratier is distorted around a professional, family man broadening his trade.

During his lifetime, Harrison was also discussed by his contemporaries as showing natural genius, or as 'the ingenious Mr Harrison'. In preparation for the trial of his first clock by the Admiralty, Harrison was discussed by the Lord High Admiral, Sir Charles Wager, in a letter to George Proctor (d.1736), who captained the *Centurion* that took Harrison to Lisbon. Wager commented that, 'the Man is said by those who knew him best, to be a very Ingenious and sober Man, and capable of finding out something more than he has already, if he can find Encouragement'.[94] In 1740, the antiquarian William Stukeley (1687–1765) described Harrison in his diary as, 'that extraordinary genius at clock-making, who bids fair for the golden prize', making an unquestioning link between Harrison's skills and the reward.[95] In 1742, Martin Folkes, with other members of the Royal Society, Oxbridge Professors, and instrument makers gave their opinion on Harrison's watches to the Board of Longitude, in which they narrated his path to working on the problems of keeping clockwork accurate at sea, saying, 'Mr Harrison at first was led, by mere Curiosity, into a Consideration of these Causes, and after various Expedients, happily contrived to remove some, and balance others against one another'.[96] They presented a picture of curious invention and 'happy', implying accidental, discovery by an uneducated man of natural ability. By 1753, in *The Inspector*, John Hill (1716–75) was making the fascinating comment that, 'it has been a sensible delight to me to hear Roubiliac extravagant on the applause of Harrison; and to see that real

92 *Annual Register* (1778), p.25. The same article also appeared in the *Universal Magazine*, *European Magazine*, and *London Review*.

93 *Annual Register* (1778), p.25.

94 Harrison, *Account*, p.18.

95 Stukeley, *Family Memoirs*, ii, p.298.

96 Reproduced as an Appendix to *Journal of the House of Commons* (12 March 1763), pp.547–8.

Genius, who knows as much of statuary as the other does of the longitude, repay the tribute with accumulation'.[97] It is interesting to see Harrison discussed as the equal of a celebrated sculptor, equating two very different forms of craftsmanship as showing comparable genius, and applauding each for the ability to recognise it in the other.

In his biography of Harrison in the 1960s, Humphrey Quill (1897–1987) emphasised Harrison's tenacity in continuing to be convinced that he could gain the great reward, and in following his dream against the disbelief of other clockmakers. He saw this as a mark of Harrison's rare genius.[98] Yet we have also seen how such stubborn or misguided chasing after (pipe) dreams was the clear mark of a projector, as was Harrison's seeming abandonment of all other work from the 1730s. Likewise, the tale of Harrison's life in the *Annual Register* ended with an apology, and an explanation for his rough style, saying,

> The recluse manner of his life in the unremitted pursuit of his favourite object, was by no means calculated to qualify him as a man of the world … In conversing on his profession, he was clear, distinct, and modest, yet, like many other meer mechanics, found a difficulty in writing; in which he adhered to a peculiar and uncouth phraseology.[99]

This used Harrison's 'peculiar and uncouth phraseology' as a means to argue for his peculiar genius. Yet this is what continually hampered Harrison's communications with the Board of Longitude and, as we will see, led them to doubt he was entirely of sound mind. Harrison could be presented as a genius, but he could also be presented as a projector or, worse, as a madman.

We have seen how new and exciting inventions trod a dangerous line on the brink of projecting, and Harrison's timekeepers balanced precariously on this boundary. Written logs were kept by the captain and lieutenants of the *Orford*, the ship that carried Harrison and his timekeeper back from the trial to Lisbon. Lieutenant Charles Knowler (fl.1734–d.1789) in his logbook tellingly described the machine that came on board as 'by the Projector John Harrison'.[100] As the fame of both Harrison and his timekeepers came to its height in the 1760s, he was discussed widely. Samuel Bamfield (fl.1764), Devonshire teacher of mathematics, put his finger on the problem in his 1764 *New Treatise of Astronomy*, when he explained that, 'as to a

97 Hill, *The Inspector*, ii, p.273.

98 Quill, *John Harrison* and *The Man who Found Longitude*.

99 *Annual Register* (1778), p.26.

100 Logs of Lieutenants Thomson and Knowler, Admiralty Papers, Caird Library, ADM/L/O/29A:49.

Time-piece, notwithstanding it is so highly improved by the ingenious and indefatigable Mr. Harrison ... none perhaps but the Inventors can, and will, execute the Workmanship with that Care and Exactness there is required for such a curious Piece of Work'.[101] We have seen that genius implied originality, but by being unreplicable, the timepieces became curious inventions, and therefore more like projecting pipe dreams. Indeed, in 1767, Harrison's inventions were compared directly to Joanna Stephens' cure for 'the stone'. In his *Letters from Altamont*, the poet Charles Jenner (1736–74) discussed how

> A certain mechanic of this inventive city, has contrived a time-piece with all the principles requisite for [finding longitude] ... There are not, however, wanting a party who say, that the people appointed to examine into the merit of the projector's claim, have been too hasty in their judgement ... it is no new thing for judgements of that kind to be given without sufficient caution. It is not many years since a reward was offered for any person who should discover a certain cure for that dreadful disorder, the stone: a woman offered her claim, and received the reward for a medicine which has since been found, in many cases, impossible to be administered, and in others ineffectual.[102]

Harrison even gained the ultimate projecting accolade, as he and his timepieces became the butt of satirical poetry. In 1771, poet and schoolmaster James Cawthorn (1719–61) published a selection of poems which included a discussion of the Gods getting drunk on Olympia, when Wit, poking malicious fun at his brother Learning,

> Ask'd him if yet his pains and care
> Has learnt to make the circle square?
> ... If his mechanic skill cou'd catch
> Perpetual motion in a watch?
> Or forge a pendulum endued
> With power to tell the longitude?[103]

Given the 1771 date, Cawthorn could not have been unaware of Harrison's work on longitude timekeepers, and Harrison must, therefore, be implicated in this typical projecting joke on squaring the circle, perpetual motion, and pendulums for finding longitude.

Such doubts and jokes about longitude in general, and Harrison more particularly, must have been known to members of the Board of Longitude, prevalent as we have seen that they were. Within Board meetings, the

101 Bamfield, *A new treatise*, p.35.

102 Jenner, *Letters from Altamont*, pp.86–8.

103 Cawthorn, *Poems*, p.195.

question of Harrison's ability to communicate his knowledge of his timekeepers was always problematic. We have seen in the previous chapter how the language used by projectors was suspect for its inflated claims and unorthodox style, and how this was extensively satirised by the Scriblerians; but also how these projectors made claims about their simple language and 'plain' reason.[104] This is certainly true of Harrison, and marks his text out as related to other contemporary genres, notably the projecting and mad writing we have been discussing. In contrast to his presentation in the newspapers and after his death, Harrison seemed consciously to distance himself from representations of his genius and unique abilities. Instead, he emphasised his hard work, long perseverance, and notably the obvious replicability of his watch. In his *Case of Mr. John Harrison* published in 1767, he argued against the great reward as being a miraculous prize like a lottery, saying, 'large as the Sum of Twenty Thousand Pounds may appear, if considered merely as the Reward of a lucky Thought or casual Discovery', this should be compared to 'the Labour of more than Forty Years of a Man capable of perfecting a Machine of such prodigious Importance'.[105] In 1772, when William Harrison wrote to the Secretary to the Treasury, John Robinson (1727–1802), in the hope that he could intercede with the Prime Minister, Lord North (1732–92), he argued that 'we should not, for the last seven Years have been employed solely in proving a self evident Truth, viz. that one machine could be made like an other … But that would probably put an End to Parliamentary Grants for Calculating … annual tables of exceeding doubtfull repute'.[106]

Yet this also quite clearly shows the projecting rhetoric absorbed by the Harrisons, in William's immediate turn to attack other proposals, in this case casting aspersions on the Board's interest in the lunar distance method. The 'Harrison Journal' makes regular, aggrieved complaints about the Commissioners' lack of attention and the responses of key figures whom the Harrisons had approached. Early on, Harrison complained bitterly of his specific treatment by the Astronomer Royal, James Bradley (1692–1762), recounting how, at one chance meeting,

> The Doctor … in the greatest passion told Mr Harrison that if it had not been for him and his plaguey Watch Mr Mayer and he should have shared Ten thousand pounds, before now. This gave Mr Harrison an Opportunity of

104 By contrast, James Amelang has looked at artisanal autobiographies of the period as showing a 'muting of self', *Flight of Icarus*, pp.123–5.

105 Harrison, *Case* (c.1767), p.3.

106 William Harrison to John Robinson, 18 December 1772, 'Letters', Harrison Papers, CLC/L/CD/E/043/MS06026/002, p.27.

> seeing what sort of a Friend Dr Bradley had been at the Board ... now self Interest seemed to be the Principle.[107]

Later, his attacks moved specifically to Nevil Maskelyne, whom the Harrisons saw as supporting the lunar distance method for his own ends, hoping for the grants from Parliament that we saw William mention above. Thus, in 1767, Harrison published his *Remarks on a Pamphlet Lately Published by the Rev. Mr Maskelyne*,[108] in which he heavily attacked Maskelyne's skills and knowledge. He complained that Maskelyne's *Account* included 'ignorant and puerile Remarks which Mr. *Maskelyne* has been suffer'd to prefix to my written Description of the Watch',[109] and that he did not doubt Maskelyne's skill in astronomy, 'But alas! As to his skill in Mechanics, he knows little or nothing of the matter he has ventur'd to take in Hand'.[110]

Harrison's greatest vitriol, and most projecting language, was preserved for the 1775 publication in which he sought to justify his entire interaction with the Board of Longitude. In this he complained at length about the Commissioners, the 'Priests and Professors', who were trying to take his rightful reward. In a footnote, he commented 'Whiston was pissed on, and Ditton shit on, but surely these Men ought to be besmear'd or bespatter'd with both'.[111] He was, therefore, clearly aware of the Scriblerians' scurrilous ode. The 'Journal' further uses exaggerated, projecting language to argue Harrison's case. In one delightfully Scriblerian outburst, he recorded saying to the Board that 'it was he that was to say which Instrument would be most proper [to be trialled]; They were to be Judges to what degree of exactness it came so that if he was to take an old Shoe and by their orders to go with it [to the West Indies] and at his return it should have been found to have answered, that certainly would be fulfilling the Act'.[112] In a letter to the Board in May 1765, he likewise commented melodramatically, 'I hope I am the first, and, for my country's sake shall be the last that suffers by pinning my faith on an English Act of Parliament'.[113] This constant return to the terms of the 1714 Act was a particular characteristic of Harrison's interaction with the Board, as we have seen, but also of pamphlets by Robert Browne and Whiston among others. In one discussion of his reliance on this Act, Harrison engaged directly with the language of credit and credibility that we discussed above, asking, 'has

---

107 *HJ*, p.12.
108 This commented on Maskelyne's *An account of the going of Mr. John Harrison's watch.*
109 Harrison, *Remarks*, p.8.
110 Harrison, *Remarks*, pp.11–12.
111 Harrison, *Description*, p.67.
112 *HJ*, p.56.
113 *HJ*, p.84.

not Mr. Harrison postponed his private Business for upwards of 40 Years, on the Credit of an Act of Parliament?'[114]

It is perhaps unsurprising, then, that the Commissioners noticeably treated Harrison like a projector on a number of occasions. After such outbursts and printed accusations in 1765, the Board chose to publish a full account of their dealings with Harrison from the first minuted meeting to the most recent. Harrison records that this decision was made 'that the world might not be imposed upon'.[115] We remember that projectors were feared because of their attempts to impose ridiculous or malicious schemes upon a trusting public. Thus, in 1766, Nevil Maskelyne published the results of his trials of Harrison's watch at Greenwich Observatory, to which we have already seen Harrison's response.[116] Harrison's anger was at Maskelyne's conclusion that the watch's rate of going was not reliable. It was also against Maskelyne's implication that Harrison knew of this irregularity, and his comment that

> These considerations are sufficient to explain the motives which might have actuated Mr. Harrison, as a man of prudence, in desiring to send his watch upon two voyages to the West Indies, upon his idea that he should be intitled to the large rewards prescribed in the act of the 12th of Queen Anne, in case his watch kept time within the limits mentioned, whether the method itself was or could be rendered generally useful and practicable, or not.[117]

He suggested, quite clearly, that Harrison was trying to make his watch appear more useful than it was by manipulating how it was seen by both the Commissioners and the public; that he was, in short, a charlatan and projector. It is clear, furthermore, that such a view was shared by at least some members of the wider public. In 1765, the *Monthly Review* had published extracts of Harrison's publications that attempted to vindicate his actions up to that point,[118] concluding that 'we cannot help being of opinion with the Commissioners, that the terms proposed to Mr. Harrison, are by no means unreasonable or improper; and cannot conceive why he should object to them; unless he has some private reason for distrusting his own discovery'.[119]

Yet Harrison did not distrust his own discovery, he believed completely in its accuracy at sea and ability to meet the terms of the great reward. It was the Commissioners that distrusted it, not just because they had no proof

114 *Account*, Harrison Papers, CLC/L/CD/E/043/MS06026/002, p.37.
115 Reported by Harrison as proposed by Hornsby, in *HJ*, p.91.
116 Maskelyne, *Account*; Harrison, *Remarks*.
117 Maskelyne, *Account*, p.23.
118 Those were *An Account* and *A Narrative*.
119 *Monthly Review* (1765), Harrison Papers, CLC/L/CD/E/041/MS03973, p.73.

that it could be made by anyone other than Harrison, but more importantly because only he had complete knowledge of it. Their increasingly desperate interactions with him suggest that the Commissioners feared Harrison was incapable of explaining his timekeepers clearly, that he was, in fact, going mad. This sense is invisible in the official minutes, but evident in the conversations that are recorded in the 'Journal'. In one exchange, this reports Lord Morton (1702–68) as describing a letter from Harrison as 'such a confused, piece of Jargon as I believe you never have heard before, and you will see from it that whoever drew it up cannot express their own minds'.[120] Such a comment immediately brings to mind Swift's discussions of the importance of expressing one's ideas in 'tune' with one's audience in order to appear sane. Harrison seemed to be aware of this problem, writing to the Commissioners, saying, 'I hope the Hon[oura]ble Board will pardon my manner of Writing and excuse my Style, as writing is not in any degree my Province … I have used my utmost efforts for the good of my fellow Creatures in the improvement of that Talent which it has pleased God to give me'.[121] We are reminded of the many other contributors who apologised for their style.

It is also noteworthy how far the Commissioners elided John and William Harrison into one person. William was always treated not merely as the spokesman for his father but almost as a ventriloquist's mouthpiece. The Harrisons did often attend meetings together, and there William would speak for his father. At the meeting on 30 May 1765, for instance, 'Mr John Harrison and Son attended the Board and being called in, Mr John Harrison told them that he was not [up] to talking, Therefore desired that his Son might speak for him, for that he had well acquainted him with his mind'.[122] Yet William was also expected to comment and swear oaths on behalf of his father without consultation. The Board essentially heard multiple voices that were conceived of and presented within the single persona of John Harrison. It is hardly surprising, therefore, that they found their interactions confused.

Such confused, we might say disordered, style is also clear in Harrison's publications. 'Mad writing' featured the constant complaint of being prevented from communicating by a controlling institution, and of having therefore, in a sense, been driven mad by the institution itself.[123] This complaint of having been prevented from communicating is one that the Harrisons made regularly. Within the same meeting where Lord Morton

120 *Monthly Review* (1765), p.89.
121 *Monthly Review* (1765), p.109.
122 *HJ*, p.80.
123 *HJ*, pp.33–7.

commented on the insane style of his letter, William was asked to sign an oath on behalf of his father:

> Mr Harrison having read it said, there is two Words in this Paper here, Lord Morton interrupted and said, Stop, Stop, Stop, Sir you were not desired to pull it to pieces, you were only desir'd to look at it and to tell us whether or not you will take that Oath, and therefore you have no Business to say anything about it only to answer to the question ask'd you.[124]

Again, at the next Board meeting, another Commissioner, Lord Egmont (1711–70), said to William, 'Sir, I have told you that we will not hear you talk, and we will not, for you are the strangest and most obstinate Creature that ever I met with'.[125] Similarly, one of the features of mad writing is being constantly questioned by doctors, and both serious and satirical treatments of Bedlam discussed the inmates being goaded by visitors. Later in the same exchange with William Harrison, Lord Egmont relented and promised that Harrison would not 'be teased by any impertinent questions'.[126]

We have already considered Harrison's use of the Scriblerian 'Ode, for Musick, on the Longitude' in his 1775 publication *A Description Concerning such Mechanism as will Afford a Nice, or True Mensuration of Time*. This pamphlet aimed to vindicate his actions and his watch, attacking the actions of the Commissioners of Longitude. As his final contribution to the longitude debate, this lengthy publication made Harrison's inability to communicate abundantly clear, as well as showing how much of his previous writing, including the 'Journal', must have been ghost written.[127] A feature of 'manic' writing of the period was its 'excessive rhetoricity',[128] a feature we have also seen of projecting pamphlets, which also perfectly describes Harrison's *Description*. This work is more than three times longer than any of his other publications, it rants and rails at the Commissioners and regularly gets lost in its own sentences. These run over multiple pages, and feature layers of footnotes and sub-footnotes. The work opens, for instance, thus:

> As first, or rather as here at the first [viz. as without the taking any Notice of the great or chief Matter, viz. of what pertains to different Vibrations, or rather, as more properly speaking, of what Advantage pertains to, or accrues from, the Largeness of a Vibration] the bare Length of a Pendulum can be no otherwise rightly considered or esteemed, but as only to what it bears, or may

124 *HJ*, pp.90–1.
125 *HJ*, p.93.
126 *HJ*, p.93.
127 See discussion of this authorship question earlier in the book.
128 Hawes, *Mania and Literary Style*, p.15.

> [as according to the common Application] bear in Proportion to the Length of the Pallats, and as together with such improper Powers or Circumstances thereunto belonging, or may, as farther thereunto belong; *i.e.* in other Words, [and as still in the first Place] ...[129]

We have not yet reached the end of the sentence and are left wondering why William did not edit this pamphlet for his father as he did the others. Harrison's language is unquestionably disordered, and this was recognised by commentators outside of the Board of Longitude. In 1775, the *Monthly Review* considered his *Description*, and commented on his language, saying,

> Any one who reads but a single page of this pamphlet will be convinced that Mr. H. is utterly unqualified to explain, by writing, his own notions, or to give a tolerable idea of his own inventions, even when he chuses to do it ... [he] deals much in abstract principles, obscure hints, and general intimations of what might be done.[130]

In accusing Harrison of focusing on 'abstract principles' and 'obscure hints', the *Review* also made the suggestion that his scheme was visionary, as well as a project. The *Description* is, indeed, partly taken up with an account of Harrison's ideas on music and his new invention for a system of musical tones.[131] In interspersing these ideas with animadversions on the Commissioners as well as his incomprehensible discussions of clockwork, Harrison reminds us forcibly of Kit Smart.

One of Harrison's most frequent accusations was that members of the Board of Longitude were after the reward for their own longitude proposals, particularly the lunar distance method, as we have seen. In a rare succinct sentence in the *Description* he commented how, 'for the love of Money, they the Professors or Priests want to prefer this [method] ... for if the Love of Money cannot be said to be the Case, they must be no better than as if out of their Senses'.[132] He summed up nicely the tension between money-seeking projects and rational or irrational thinking that featured in discussions of the longitude problem, and that dogged his own interactions with the Board. Harrison himself acknowledged these discussions and tensions at another point in the *Description*, when he commented in a footnote,

129 Harrison, *Description*, pp.3–4.

130 *Monthly Review* (1775), Harrison Papers, CLC/L/CD/E/041/MS03973, p.320. Harrison responded in an appendix to the *Description* published later in 1775, in which he called the review 'Nonsense, Spite and Poison, scandalously scurvy, dirty Work', p.114.

131 Harrison, *Description*, pp.68–100.

132 Harrison, *Description*, p.61

> Their Cambridge and Oxford Mechanicks ... have writ and published a great deal of Stuff, pertaining, as they think, to the Longitude, or at least would have others so to believe it to be; whenas it is still no farther than as what Mr. Whiston did ... the which only made or could make a great Noise, and the Greatness or Impertinence or their Superfluity, as not having any Thing to do in the Matter ... being only a deal of Stuff, so as whereby to dazzle the World.[133]

He crowed in another footnote that 'their ingenuous or disingenuous Villainy, being at the Height ... fell, and indeed it was a very great Fall, the Matter being got so excessively high'.[134] Harrison quite clearly referenced the imagery of bubbles, over-inflated ideas, and dazzling projects, which characterised the language used to discuss the longitude problem and brought this full circle onto the Commissioners.

'Of the Longitude Hoard / Which is rul'd by the Board'

By 1775, it seems, the whole longitude debate had become itself a bubble, a space in which accusations of projecting and madness were levelled by all sides and were made using the same language of ephemerality and credit. Harrison was one of the more colourful contributors. In this, he was rivalled only by one other commentator on the debate, the prolific publisher of natural philosophical and mathematical pamphlets Robert Heath (fl.1737–d.1779). We have already seen Heath joining the chorus of satire on William Whiston, in his *Natural and Historical Account of the Islands of Scilly*, but his vitriol was extensive, allowing him to accuse any and all contributors to the debate of either projecting or madness, or both. It is clear that by the time that he ran out of steam, Heath considered the entire endeavour of longitude—Board, reward, proposals, and all—as a mad project. One of his first publications in this vein, *The Palladium-Supplement* of 1764, was dedicated in support to the Commissioners of Longitude, and set up mathematics as the source of reason against Harrison as projector. Heath thus opened with an argument in support of mathematicians, saying,

> Where the Mind works on *mathematical Ideas*, it works securely; which cannot be done in other *Things* so truly, because one cannot keep so strictly to the *Definitions*, or Meaning of Words in other Subjects; where the *Ideas* are often

133 Harrison, *Description*, pp.48–9 note. Harrison's accusations clearly upset the Commissioners, as shown by a draft amendment to the 1765 Act stating their ineligibility for the reward money. See Barrett, '"Explaining" Themselves', pp.152, 158.

134 Harrison, *Description*, p.33 note.

> confounded … therefore *mathematical Demonstrations* are the most proper to cleanse the Mind from *Error.*[135]

This was followed by an essay on 'The Insufficiency of Mr. Harrison's Watch', likewise addressed to the Commissioners, which argued that,

> Though … we wish to see Mr. *Harrison* duly rewarded … we find there have ever appeared more *Pretenders* than Persons of *real* Ability to discover the Longitude, whereby the *honourable Persons* appointed to judge of the Methods proposed for that *End*, are … liable to be imposed on by *specious Pretences.*[136]

Heath thus not only drew on precisely the discourses of projecting and madness discussed above—noticeably using the same political allusion to the danger of 'pretenders' that we have seen Swift used in his letter to Arbuthnot—but also anticipated the concerns that Maskelyne would articulate in his *Account*. Heath termed the public interest in Harrison 'the Grand Delusion of this *Time-piece*' and went on to comment on the 'vain *Boasts*, *Signatures*, and *Pretensions*' that characterised Harrison's publications.[137]

This view continued up to 1768, when Heath published his *Supplement to the Royal Astronomer and Navigator*, in which he commended Maskelyne's trials for exposing Harrison's watch as a fraud and made a topical projecting comparison. He argued that 'our Astronomer Royal, has infallibly proved, by a *Series* of correct Observations … that it can, *by no Means*, answer the *End proposed*, the Proof thereof … like that of Mrs. *Stevens*'s *Medicine*, was made *too late*'.[138] Yet in the same year, Heath included a 'Longitude Ode, by Mr Moonsby' in the *British Palladium*, which owed an obvious debt to the Scriblerians. This gave a brief history of the longitude problem, attacking most participants:

> I. Disputes still arise,
> For the *Longitude Prize*,
> Since *Whiston* and *Ditton* are fled;
> And H[ar]r[iso]n's W[atc]h,
> Has proved a mere *Catch*,
> And goes like one out of it's *Head*, it's *Head*,
> And goes like one out of it's *Head*. (See Mr. Maskelyne's
> *Observations*) …

135 Heath, *Palladium-Supplement*, p.iii.
136 Heath, *Palladium-Supplement*, pp.15–16.
137 Heath, *Palladium-Supplement*, pp.16–17, 20.
138 Heath, *Supplement to the Royal Astronomer*, p.1.

VI. Of the Longitude *Hoard*,
Which is rul'd by the *Board*,
No *Em[e]rs[o]n*, ever yet shar'd;
And the *Nautical Nac*,
Is but a fam'd *Crack*,
Where a *Halley* yet never *appear'd*, *appear'd*,
Where a *Halley* yet never *appear'd*.
VII. Of *Cambridge* and *Lyon*,
And *Oxford*, cry fye on!
No Longitude yet has been found;
The learned Professors,
Have all been Aggressors,
And *M[a]sk[e]ly-ne's* only *renown'd*, *renown'd*,
And *M[a]sk[e]ly-ne's* only *renown'd*!'[139]

We note the distinctly sarcastic note to Maskelyne's 'renown'd' status here, as well as a critique of the *Nautical Almanac* as 'cracked'.

In 1770, Heath started to publish *The Seaman's Guide to the Longitude: or a Key to the Nautical Almanac*, which was now addressed to the Commissioners as 'the Connoisseurs of Longitude'.[140] This complained that Maskelyne had still not published Mayer's tables, and accused him, like Harrison, of seeking the reward for himself.[141] This time, it was Maskelyne who was compared to a medical projector, as Heath commented that 'the Variety of intricate Rules and Methods used in teaching the Use of the *Nautical Ephemeris* … may be compared to the *Nostrums* of the *Quack-Doctors*'.[142] In his 1773 *British Palladium*, he included a long poem entitled 'The Longitude, A Song' of which we will hear more in the next chapter. This listed Harrison and Maskelyne as projectors side by side:

As Harrison's Watch by the Stars has been try'd,
Which, for not going true, is by them set aside;
So the Longitude-Book by the Sun, Moon, and Stars,
Should be prov'd – that their Distances prove not a Farce.[143]

139 Heath, *British Palladium* (1768), p.72. Heath's support for Emerson is also noteworthy. Emerson was a regular contributor to both the *Ladies Diary* and *British Palladium*, and was supported by Heath in a controversy in the 1740s. He published best-selling practical works on astronomy and mathematics, precisely what Heath argued were needed for ordinary seaman.

140 Heath, *Seaman's Guide*.

141 Mayer's tables would be published not long after as *Tabulæ motuum*.

142 Heath, *Seaman's Guide*, p.42.

143 Heath, *British Palladium* (1773), p.2.

By 1775, Heath was referring to Maskelyne as a 'longitudinarian'.[144] Thus, the language used across contemporary discussions to consider longitude as a project or mad scheme, which so clearly affected the interactions of Harrison and the Commissioners of Longitude, could be applied equally to both. It cast doubt on the whole epistemological space of the longitude problem, making a 'longitudinarian' in natural philosophy as dangerous as a 'latitudinarian' in religion.

144 Heath, *British Palladium* (1775), p.2.

Conclusion to Part II

Accusations of projecting and madness were also levelled at Hogarth. Fellow artist Paul Sandby (1731–1809) produced an entire set of satirical prints in 1753, later named *The Analysis of Deformity*, which attacked Hogarth's book *The Analysis of Beauty*, presenting him as insane, a projector foisting dangerous ideas onto the public, using the full range of imagery that we have considered above.[1] The title image to the set, *The New Dunciad*, made obvious reference to Pope's poetic attack on projecting literature.[2] In the image, Sandby separated Hogarth from a comical harlequin figure behind him, labelled 'his genius', and represented his entire project as an unstable house of cards. In another, he presented Hogarth as a mountebank, 'this arrogant quacking analyst', selling his 'cure' for art from a stage surrounded by foolish investors.[3] In *The Analyst Besh—n in his own Taste*, Sandby used the same scatological satire that the Scriblerians had directed against Whiston and Ditton, and which Harrison had redirected at the Commissioners.[4] He likewise represented Hogarth's *Analysis* being 'thrown into the Caves of Dulness and Oblivion', clearly referencing Pope's commentary on the homes of projecting literature and mathematical madness. The series culminated

1 These are analysed and placed within Sandby's broader *oeuvre* in Quilley, 'The Analysis of Deceit', pp.38–47.

2 Paul Sandby, *The New Dunciad* (1753); etching, 390 x 290mm, Lewis Walpole Library.

3 Paul Sandby, *A Mountebank Painter* (1753); etching, 232 x 192mm, British Museum.

4 Paul Sandby, *The Analyst Besh—n in his own Taste* (1753); etching, 263 x 182mm, Lewis Walpole Library.

Figure 52: Paul Sandby, *The Author Run Mad* (1753); etching, 270 x 180mm, courtesy of the Lewis Walpole Library, Yale University.

in *The Author Run Mad* (Figure 52), in which Sandby represented Hogarth in a direct satire on his image of Bedlam. Sandby turns Hogarth into our longitude lunatic, portraying him scribbling ideas on the wall, chained within a circle on the floor, his 'line of beauty' become a mark of his delusion. He shows us how tightly imagery of madness and projecting had become bound up with longitude, putting it at the centre of the full spectrum of contemporary satire to comment on Hogarth's artistic aspirations as a mad project. To make Hogarth mad, what better than to turn him into his own longitude lunatic 'projecting' his images onto the wall?[5]

In his *Autobiographical Notes* in response to these criticisms, Hogarth compared himself to a watchmaker, perhaps to Harrison, arguing that a watchmaker 'may say the [watch] I have made I'll warrant you is as good [as] any other man can make you & if it … really is so the watch maker is not Branded as Infamous, but on the contrary is esteem[ed] as an honest man who is a[s] good as his word'.[6] He compared this directly to his own treatment by his fellow artists when he attempted to theorise with them. This was, however, precisely John Harrison's problem, of how to communicate his practical knowledge to the Board of Longitude in such a way that he would be 'esteem[ed] as an honest man who is a[s] good as his word'. It was both his honesty and the goodness of his word that were questioned: his ability to communicate with the Board in the language that they would understand, such that they would see him neither as a commercialised projector, nor as a disordered madman. Thinking back to Harrison's engraved portrait in comparison to John Bird's, we note that his words are not visible on the printed page, whereas Bird's are (Figures 39–42). His words were either untrustworthy or unreasonable, or both.

5 Lynall suggests 'projecting's' relationship to images in 'Scriblerian Projections', p.1.
6 'Autobiographical Notes' in *Analysis*, pp.217–18.

III

The Social Problem

'Experimental Exhibitions'

In whatever character we examine the affairs of human life, *man* may be considered … as a *time-keeper*, not free of imperfections. So far as intellectual and mechanical powers can be compared, it is by the use of his *time*, the *longitude and latitude* of all his thoughts, words, and actions are to be determined.[1]

1 Hanway, *Thoughts*, p.51.

Figure 53:
William Hogarth,
A Rake's Progress,
Plate 8 (1763);
engraving,
396 x 478mm,
gift of Sarah
Lazarus, 1891,
Metropolitan
Museum of Art,
New York.

5

Polite or Impolite Science?

By the early 1760s, Hogarth was becoming increasingly disillusioned with society. From affectionate but ribald commentaries on his contemporaries, his satires became darker and more actively critical. In 1763, he re-worked the series of *A Rake's Progress* engravings, making distinctive changes to his Bedlam scene (Figure 53): a half-penny appeared on the wall partly covering the longitude diagram, and the two women in the background changed their dress, posture, and expressions. These women are the only figures in Bedlam that engage with the viewer, gazing out from amongst the madmen, all the more so in the 1763 re-working. Their presence in this setting establishes them as two types of viewer of both artistic and natural philosophical display: they are both a mark of refinement amongst the disorder,[2] and they are delighted voyeurs. They look intently both at the lunatic's considered public display of his longitude drawing, and at the half-naked body that he also exhibits. The previous chapters have all carried an underlying theme of social self-display as different longitude actors sought to present both their images and their language in such a way that it appealed both to the public and to the Board of Longitude. We have looked at the visual and mental questions that the longitude problem negotiated. Here we turn our attention to social questions, and to what this meant for these actors and for John Harrison.

The eighteenth century showed a particularly 'exuberant exhibitionism' as a product precisely of the political, religious, and market forces discussed

2 Solkin, *Painting for Money*, pp.78–105.

in the previous chapters, an exhibitionism which itself became the subject of comment and satire.[3] Many writers saw an inherent tension between the polite and the impolite in society. The attempt to be polite required performance and effort, knowledge of what you could and could not do, say, or display in specific contexts, and the concealment in plain sight of acknowledged but rejected aspects of men and women in society. The longitude problem was negotiated in such social settings, raising questions of concealment, display, and social performance for our actors. Thus, longitude discussions took place within the parameters of 'polite' science but also on its edges, used in fact to question and articulate where those boundaries lay. The concern of behaving properly in different settings affected John Harrison's performance as much as his language, allowing us to understand the depth of social anxiety behind his complaint to the Commissioners in the 1760s that 'he might never know when he had done from the words, Experimental Exhibitions'.[4]

Longitude appears in a surprising variety of works that discuss social relationships, forming part of the exploration and criticism of the practices that formed society. In Conolly's (fl.1736) 1736 play *The Connoisseur*, while defending his right to be interested in curiosities, the protagonist argued, 'Amusement is the great Secret of Life – it's … the Longitude, the perpetual Motion – in a Word, it's Pleasure without Pain'.[5] He equated longitude directly with the constant search for diversion, pleasure, and amusement. One site in which this search was pursued was Bedlam itself, infamously open to visitors in the eighteenth century as part of the tour of sociable display that also included sights such as the crown jewels and the animal menagerie at the Tower, the scientific instrument collections at Gresham College, and the entertainments at Vauxhall Gardens. The Scriblerians made great play with this, clearly linking Bedlam to sites of natural philosophical investigation and display in Swift's Academy of Lagado. Similarly, Swift wrote to his companion Esther Johnson (1681–1728), 'Stella', in December 1710 how he himself had seen 'all the sights, lions, &c. then to Bedlam; then dined at the Chop-house behind the Exchange; then to Gresham College … and concluded the night at the Puppet-shew'.[6] Bedlam was a place of sexual as well as mental licence, a transgressive space in which public codes of behaviour were articulated and resisted.[7] In his *London Spy* account

3 Agnew, *Worlds Apart*, pp.1–16, quotations pp.13, 14. This also relates to Solkin's work on changing attitudes to art as a commodity in *Painting for Money*.

4 *HJ*, p.83.

5 Conolly, *The Connoisseur*, p.69.

6 Swift, *Journal to Stella*, i, pp.122–3.

7 Stallybrass and White, *Politics and Poetics*.

of Bedlam, which we have already encountered, Ned Ward commented on 'the Looseness of the Spectators, amongst whom we observ'd abundance of Intriguing; Mistresses, we found, were to be had of all Ranks, Qualities, Colours, Prices, and Sizes ... Commodities of all sorts went off, for there wanted not a Suitable *Jack* to every *Jill*'.[8]

Masquerades were another such transgressive London space in which to see and be seen, particularly popular in the 1720s and '30s and re-appearing in the 1760s. In 1728, Henry Fielding (1707–54), author and theatrical entrepreneur, published a poem entitled *The Masquerade*, supposedly by Swift's Lemuel Gulliver, in which he discussed the motives of fashionable society at a masquerade and linked these directly to the longitude problem.[9] He opened by arguing that

> Some call Curiosity an evil,
> And say 'twas that, by which the devil
> With Eve succeeded in his suit,
> To taste the dear forbidden fruit.
> Others (allowing this) yet wou'd
> Prove it has done less harm than good.
> To this (they say) whate'er we know
> In arts or sciences, we owe.
> To this, how justly, are attributed
> What W[hi]st[o]n, H[al]l[e]y, have exhibited!
> From this we borrow hopes of greater
> Discoveries of madam Nature.
> Hence is our expectation gain'd,
> To see the longitude explain'd ...
> By this embolden'd, tim'rous maids
> Adventure to the masquerades.[10]

He moved from curiosity as an inroad to sin, to seeing it as the impetus behind work on the longitude 'exhibited' by Whiston and Halley, and back to a motive for young women to attend such shady gatherings. He went on to emphasise the mixing and upsetting of social norms in these spaces—'here in one confusion hurl'd, / Seem all the nations of the world; / Cardinals, quakers, judges dance ... Who masque the face, t' unmasque the mind'[11]—and how the event was itself a mask for intrigue and sexual licence: 'They'll

8 *London Spy*, p.67.

9 Arbuthnot, *Miscellaneous Works* attributed this poem to him, linking it into the satirical traditions discussed in Chapter 3.

10 Fielding, *Masquerade*, pp.1–2.

11 Fielding, *Masquerade*, pp.2–3.

tell you virtue is a masque: / But it wou'd look extremely queer / In any one, to wear it here'.[12] Thus, from an early stage, the longitude problem was referenced in spaces within which polite social appearance conflicted and blended with its darker impolite undertones.

A Polite Science

'Politeness' has become an established concept for understanding the eighteenth century. Seen as the framing ideology of the 'public sphere', politeness was part of a widening of popular involvement in knowledge exchange that was focused in the newly emerging social spaces, in coffee houses, taverns, and salons, and in the new literary forms, such as newspapers and novels. The 'public sphere' was the force that animated the 'Enlightenment' and led to wide political, economic, and social changes by the end of the century.[13] The liberal arts have, further, been seen as the central tool of such politeness, through the 'pleasures of the imagination' that allowed refined conversation, social interaction, and living spaces.[14] Natural philosophy, and particularly astronomy, was the perfect subject for this public sphere, a subject that could be displayed and admired in lectures, displays, and elegant instruments, and could be discussed between the sexes in witty conversation.[15] It provided the space for women to improve men's morals and forms of social interaction, while men improved female minds, focusing them away from the dangerous spaces of public entertainment, like masquerades.[16] In 1755, one public lecturer in natural philosophy, Stephen Demainbray (1710–1782), even advertised that every gentleman attending a lecture would receive a free ticket for a lady.[17] Natural philosophy was part of a culture of performance and theatre.[18] The longitude problem linked particularly well into polite science with its focus on astronomy, mapping, and instrumentation. Three aspects of polite science particularly featured: the

12 Fielding, *Masquerade*, p.6.

13 Habermas, *Structural Transformation*.

14 Brewer, *Pleasures of the Imagination*.

15 Broman, 'The Habermasian Public Sphere and "Science" in the Enlightenment', pp.301–36 and Secord, 'How Scientific Conversation became Shop Talk', pp.129–56. Walters argues particularly for the role of astronomy in 'Conversation Pieces', pp.121–54 and 'Tools of Enlightenment'.

16 Walters, 'Conversation Pieces'. Such polite men and women are also those most likely to suffer from the maladies discussed by Cheyne in Chapter 4.

17 Morton, 'Lectures on Natural Philosophy', p.420.

18 Schaffer, 'Natural Philosophy', pp.1–43.

social status of instruments, the emphasis on image and appearance, and the role of conversation.[19] All three required performance to be polite.

INSTRUMENTS

IN 1710, the Frankfurt collector and traveller Zacharius Conrad von Uffenbach (1683–1734) made an extended stay in London, visiting instrument shops, key figures, and natural philosophical sites amongst coffee houses, plays, and dinners, giving us a view of the world that would soon be galvanised by the Longitude Act. Uffenbach placed particular emphasis on the appearance of both the men and machines that he encountered. He commented on 'the elegance, care and neatness of the workmanship' of the new turret clock in St Paul's Cathedral,[20] compared with the disgusting state of the collections at Gresham College, where 'the finest instruments and other articles ... [are] not only in no sort of order or tidiness but covered with dust, filth and coal-smoke, and many of them broken and utterly ruined',[21] or with the conditions at the Greenwich Observatory, 'far more wretched both as regards comfort and equipment than we had imagined'.[22] Yet on this same visit he also commented favourably of Flamsteed that 'he received us with uncommon civility (which is in England most unusual)'.[23] Similarly, on visiting the workshop of instrument maker Francis Hawksbee (1660–1713), 'we saw with amazement his ... demonstrations and experiments ... very excellent and curious. We took up the book of experiments he has published and got him to show us one after another'.[24] Uffenbach's visit tied Hawksbee's instruments, publications, and experimental displays into a single polite encounter. This compared markedly to his reception by the collector and Fellow of the Royal Society John Woodward (1665–1728), the butt of so much Scriblerian humour,[25] who 'showed us all his things with such affected gestures and rolling eyes that we could not restrain our laughter ... he behaves like a female and an insolent fool'.[26] We will return to such behavioural, gendered concerns in Chapter 6.

19 Florence Grant has considered the material qualities of a particular collection of instruments within these broader cultural parameters in 'Philosophical Instruments'.
20 Uffenbach, *London*, p.33.
21 Uffenbach, *London*, p.98.
22 Uffenbach, *London*, p.22.
23 Uffenbach, *London*, p.21.
24 Uffenbach, *London*, pp.168–9.
25 Kirby-Miller, *Memoirs*, pp.102–4.
26 Uffenbach, *London*, p.178.

Those who could not visit notable collections in person could attend public lectures and experiments. These figured in the social calendar alongside other types of public amusement, including fireworks and 'naumachia' (floating entertainments), and these are precisely the genres of display on which longitude proposers drew.[27] Whiston and Ditton's 1714 proposal placed flamboyant public display at the heart of the process.[28] Their idea for using shells had been inspired by the celebratory fireworks of 1713,[29] which they proposed to replicate by setting off fireworks on Hampstead Heath for the public to observe and contribute their calculations (as we heard satirised in Chapter 4).[30] An annotated advertisement in the Royal Observatory archives shows that at least one reader did answer their request, but then wrongly sent their observations to Flamsteed.[31] Whiston continued this engagement in his later attempts, including lists of observations to be carried out by his readers in his 1724 publication on solar eclipses,[32] and in 1741 proposing using 'balls of fire' to correct Halley's coastal charts.[33] The elaborate structures required to stage lavish public firework displays had spread to England from Italy, bringing with them their Italian description as *macchine*. The fireworks staged in Green Park to celebrate the end of the War of the Austrian Succession in 1749 were the most famous and most controversial, criticised by many commentators as a frivolous waste of money. Yet in 1742 Benjamin Robins (1707–51), a Fellow of the Royal Society, had published *New Principles of Gunnery* arguing that projectiles offered serious experimental possibilities (no matter the arguments of Whiston satirists that we saw in Chapter 4). It is possible that Robins was, likewise, the author of an article in the *Gentleman's Magazine* in 1748 that suggested the Green Park fireworks as an opportunity for 'ingenious gentlemen' to make observations.[34] Thus, fireworks continued to offer both practical and exhibitionist opportunities. In presenting the original use of the term *macchine* or 'machine' they, likewise, linked forwards to Harrison's famous 'machine' H4, and the sorts of display that would be

27 Altick, *Shows of London*, pp.87–98.

28 John Loftis has discussed Whiston's role as a public lecturer at Richard Steele's Censorium, showing how he was accustomed to such public showmanship, in 'Richard Steele's Censorium', pp.43–66.

29 Whiston, *New method for discovering the longitude*, pp.22–3.

30 *Post Man*, 14 July 1715 and 14 May 1717; *Daily Courant*, 5 June 1717.

31 'Guardbook', Royal Greenwich Observatory Archives, RGO 1/37: 150.

32 Whiston, *Calculation of solar eclipses*, pp.76–8.

33 Whiston, *Longitude Discovered by the eclipses* (1741), pp.xxxv–vii, quotation p.xxxv.

34 Werrett, *Fireworks*, pp.143–6, 149–55; Robins, *New principles*; 'A Geometrical Use Proposed from the Fire-works' in the *Gentleman's Magazine* 18 (November 1748), p.488.

required of him. Such lights in the sky of course also link back to Whiston's prophetic ideas discussed in the previous chapter.[35]

Two decades later, in the 1760s, two foreign natural philosophers and diplomats visited London instrument makers and natural philosophers actively engaged with the longitude problem, as part of their attempts to obtain up-to-date navigational knowledge for their respective monarchs, surreptitiously if necessary. Yet the Swedish Professor Bengt Ferrner (1724–1802) and the French socialite and editor of the *Connoissance des Temps*, Jérôme Lalande (1732–1807), also enjoyed the more varied entertainments offered by London. Ferrner visited Covent Garden and Drury Lane and paid to see fairground marvels, as well as scientific displays like Edward Nairne's (1726–1806) electrical experiments. Lalande enjoyed the Tower, the British Museum, and an execution at Tyburn, as well as the pleasure gardens at Ranelagh and Vauxhall 'where time passed deliciously'.[36] Both also spent time in taverns socialising with instrument makers and joined the Royal Society Club meetings. Ferrner attended the George and Vertue Tavern with a group of instrument makers, where he met Short, Bird, and Dollond, and ordered instruments for the Stockholm observatory.[37] On 3 March 1760, he commented, 'I spent all day at Short's, Dollond's and Bird's, and entertained myself by seeing their exquisite pieces of work'.[38] Lalande befriended members of the Board of Longitude itself, discussing Harrison's clocks with Lord Morton a number of times. Both men visited Christopher Irwin's chair, which Jeremiah Sisson had mounted on the roof of his house, Ferrner describing it, tellingly, as 'polite and comfortable'.[39] Both also went to Greenwich, where Ferrner dined with Bradley at the Observatory and discussed various longitude solutions.[40] What both Ferrner and Lalande really wanted were the details of Harrison's famous timekeepers, to which we will return.

IMAGERY

INSTRUMENTS AND NATURAL PHILOSOPHY did not only perform physically in the polite social space, but also through illustrations in publications, newspapers, trade cards, and prints.[41] Images like the print

35 Werrett, *Fireworks*, pp.96–8, 149–50.

36 Lalande, *Diary*, pp.33–4.

37 Orrje, 'Navigating the Neighbourhoods', pp.27–30; Lalande, *Diary*, pp.9–10.

38 Ferrner, *Resa i Europa*, p.216. My thanks to Jacob Orrje for these translations.

39 Orrje, 'Navigating the Neighbourhoods', pp.24–5.

40 Orrje, 'Navigating the Neighbourhoods', p.32.

41 Walters argues that illustrated advertisements were most powerful and acted like

Astronomy dating from the 1740s (Figure 54) show the ideal of men and women engaged in polite interaction through the use and discussion of instruments in classical settings.[42] This was the type of interaction encouraged by the *Universal Magazine of Knowledge and Pleasure*, published from 1747, which was in many ways an English precursor to the famous French *Encyclopédie*, featuring topics ranging from famous figures, industrial processes, artists, and artistic techniques, to games, dances, and fashion. Each volume featured a different frontispiece emphasising the polite interaction between the arts and sciences that the magazine envisioned. Those from 1747 (Figure 55) and 1755 (Figure 56) are illustrative in placing men and women discussing natural philosophy in a library and garden, respectively. Indeed, the 1755 frontispiece is copied directly from the *Astronomy* print (Figure 54). Articles included discussions of many different types of instruments, and even the occasional advertising plug for a specific maker's model.[43] Timepieces seemed to be worthy of particular interest, with observations on clockwork appearing from the very first issue in 1747. In 1748, a reply to these by 'Philo-Chronos' discussed at length *The Art of Making Clocks and Watches*, hoping to give 'a Description of an Art, which, though its Principles are but little known, is carried in the Pocket of every *Gentleman*, and become the genteelest Piece of Furniture in almost every Cottage'.[44] This was accompanied by an engraved illustration showing a polite be-wigged man, sat on an elegant chair in a well-lit, panelled room (Figure 57). His tools are carefully arranged on the table and two examples of his art are displayed against the walls. He also exhibits two diagrams illustrating the intricacies of clockwork as explained by Philo-Chronos, where that on the left directly copies William Derham's *Artificial Clock-Maker* discussed in Chapter 2 (Figure 33). Thus, Derham's work to improve clockwork for use at sea was displayed on the wall of this polite workshop for all readers of the *Universal Magazine* to enjoy.[45] Later

street signs, see 'Tools of Enlightenment', pp.139–45. For some specific examples of illustrated images at work in the period see Bryden, 'Figurative Instruments', pp.2–10.

42 Snodin has discussed how such prints were made possible by the Engravers' Act in *Rococo*, p.32. Dunn discusses how the engraving is probably copied from the work of Jacopo Amigoni, who was active in England in the 1720s and '30s, as well as how the image spread to glass painting and even tiles. See 'Taming the Telescope', p.197.

43 Walters, 'Tools of Enlightenment', pp.103–8.

44 *Universal Magazine* (January–June 1748), p.88.

45 Jonathan Betts and Celina Fox have noted that many of the engravings for the magazine were done by Charles Grignion, son and brother to the clockmakers Daniel and Thomas Grignion, so that we may here see a representation of their workshop. Fox also suggests that the young man may be Thomas. Personal communication from Jonathan Betts; Fox, *The Arts of Industry*, p.254. Grignion also made many engravings after Hogarth, as well as the frontispiece to Martin's *Young Gentleman and Lady's Philosophy*.

Figure 54: Anon., *Astronomy* (c.1740); etching, 270 x 375mm, Royal Museums Greenwich, London.

Figure 55: Frontispiece to the *Universal Magazine of Knowledge and Pleasure* (London, 1747); etching, 215 x 124mm, reproduced by kind permission of the Syndics of Cambridge University Library [T900.d.75.1].

Figure 56: Frontispiece to the *Universal Magazine of Knowledge and Pleasure* (London, 1755); etching, 194 x 105mm, Private Collection.

Figure 57: Anon. (possibly Charles Grignion), 'The Art of Making Clocks and Watches' from the *Universal Magazine* (London, 1748); etching, 183 x 223mm, Wellcome Collection, Public Domain.

Figure 58: Anon., 'The first Lecture in the Sciences of Geography and Astronomy' from the *Universal Magazine* (London, 1748); etching, 216 x 262mm, reproduced by kind permission of the Syndics of Cambridge University Library [T900.d.75.2].

issues of the *Magazine* went on to feature courses of lectures on geography and astronomy, and then on experimental philosophy, both complete with an engraved image (Figure 58). The geographical lectures opened by considering the problems of drawing a line of longitude—'Behold the sphere, and observe the several imaginary circles described on the artificial globe'[46]—and explained how and why longitude and latitude lines worked. Instrument makers such as Benjamin Martin (1705–82) and Adam Walker (1730/31–1821) gave just such lectures from the 1740s to 1770s.[47] In 1752, one Owen Boscawen (fl.1752) even submitted his 'Essay for finding the Longitude' to the *Magazine*, arguing that all that was needed was an hourglass and a plumb line to observe when a fixed star crossed the meridian.[48] This was published, but seems to have provoked no further discussion.[49] In 1759, another article discussed magnetic variation as a means to find longitude, referencing Halley's maps as well as Irwin's chair and the method of using Jupiter's satellites.[50]

At least one specific longitude pamphleteer embraced this genre of image to showcase his instrument in a polite setting. In 1740, John Barston (fl.1738–40) published *The Description and Uses of Barston's New-Invented Universal Astronomical Quadrant*, explaining how he had improved on the existing quadrants by Hadley and others. He argued that 'the manner of using it is so obvious, that it is almost impossible not to apprehend it at first sight'.[51] His image (Figure 59) worked to endorse this claim, showing his quadrant in use by fashionably dressed gentlemen at their ease in outdoor spaces from idyllic parkland to a ship at sea. They use the quadrant either hand-held or on two different pedestals, as Barston instructs. This relates directly to the gentlemen and ladies peering through telescopes in *Astronomy* (Figure 54), and to the gentleman who demonstrates using a quadrant for the *Universal Magazine*'s 1752 article on mathematical instruments (Figure 60). Both turn their quadrant into a sculptural object on a pedestal, as much an object at which to look, as with which to look. We remember Benjamin Cole's striking image in Chapter 2 of his quadrant being used by a man squashed in between

46 *Universal Magazine* 2 (January–June 1748), p.285.

47 See Walker, *Syllabus* and Martin, *Course of Lectures*. John Millburn has discussed Martin's early itinerant lectures in the 1740s and '50s in *Benjamin Martin*, pp.35–64.

48 Solutions were also sent to other fashionable newspapers like the *Gentleman's Magazine*, such as in October 1737, vii, pp.615–16, and May 1738, viii, p.679 (misnumbered, should read '279').

49 *Universal Magazine* 11 (July–December 1752), pp.85–7.

50 'Observations on the Variation of the Needle; with an Hypothesis for discovering and easily knowing the longitude' in *Universal Magazine* 25 (July–December 1759), pp.173–4.

51 *Description and Uses of Barston's*, p.6.

Figure 59: Figure from *The Description and Uses of Barston's New-Invented Universal Astronomical Quadrant* (London, 1740); engraving, 202 x 257mm, Bodleian Library, University of Oxford: G.Pamph. 2063 (1).

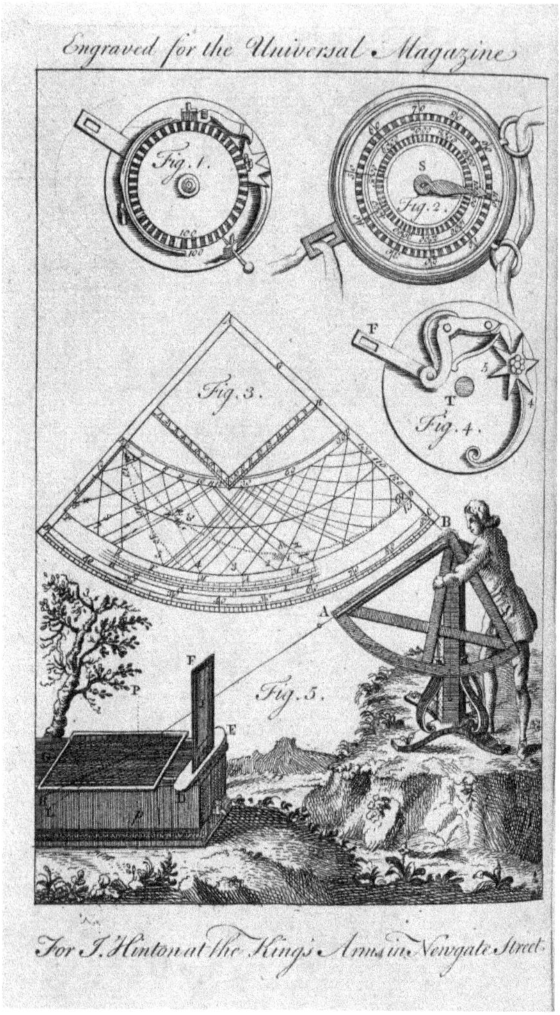

Figure 60: Anon., 'The Construction and Use of several Mathematical Instruments' in the *Universal Magazine* (London, 1752); etching, 216 x 125mm, reproduced by kind permission of the Syndics of Cambridge University Library [T900.d.75.11].

the backstaff and Hadley's quadrant. In his pamphlet, Barston in fact actively compared his quadrant with other famous instruments that would have been illustrated and visited, saying,

> No wonder, then, if *Tycho Brahe* and the rest of the Ancients differed so much from the truth in these, when 'tis plain the fault lay in their Instruments and not in their Skill ... they were at great Expence, made large Instruments, and when done, could not come near the Mark. But now by this portable Quadrant, we are enabled to measure the great Distances in the Heavens that come within the reach of our Sight.[52]

52 *Description and Uses of Barston's*, p.27.

Figure 61: 'Mathematicks', from John Savage, after François Boitard, Instrument Cards (1740); engraving, 91 x 57mm, © Trustees of the British Museum.

By comparing his instrument to those of Tycho Brahe (1546–1601), he not only argued for its accuracy compared to Tycho's known precision, but also for its equal status as an instrument of polite interest to Brahe's large expensive ones experienced in a courtly setting.

Such instruments and images also featured on wider printed materials. In 1700, John Savage (fl.1683–1701) and Thomas Tuttell (c.1674–1702) produced a set of playing cards which showcased different scientific instruments and trades,[53] linking back, of course, to the bubble cards discussed in Chapter 3 (Figure 50). Savage's cards showed instruments in a range of domestic and commercial settings, comparable to those endorsed by the *Universal Magazine*. 'Mathematicks' showed a group of men in a tiled and panelled room, conversing about instruments which, according to the accompanying text, 'by the assistance of a Master are made so delightfull & pleasant, & of such general use, that they aggrandize a Man for all Conversation' (Figure 61). A large number of other cards particularly featured navigational instruments, including a cross staff, a telescope, four different quadrants, a compass, two

53 Fara has discussed the card among these that featured the magnet in the context of polite science in 'A Treasure of Hidden Vertues', pp.5–35.

Figure 62: John Bird, Trade Card (c.1760s); engraving, 147 x 92mm, courtesy of the Lewis Walpole Library, Yale University.

Figure 63: R. Rust, Trade Card (1740s); etching, 166 x 126mm, © Trustees of the British Museum.

types of globe, charts, and an almanac. There are also specific cards for a 'Shipwright', for a 'Shipp' as an item of 'Art', and one listing navigational books and instruments. We can imagine a game of whist after supper requiring gentlemen and ladies to pay attention to these navigational tools. In 1730, longitude featured in a game for telling fortunes entitled *Dr. Flamstead's and Mr. Patridge's New Fortune-Book*, which used cards to answer a series of questions. There are separate answers to each for a man and a woman. To the question 'Whether you shall keep a coach or chariot?' the two answers come:

> A Coach! a Wheelbarrow you mean,
> For such a saucy upstart Queen
> [or]
> He will a Chariot keep, no doubt,
> When he the *Longitude* finds out.[54]

54 Partridge and Flamsted, *Dr. Flamstead's*, p.59. This second answer echoes interestingly the wording of *The Star-gazer* considered below.

Figure 64: Thomas Blunt, Trade Card (1790s); etching, 86 x 132mm, © Trustees of the British Museum.

Another type of card that entered and in turn pictured polite social spaces was the trade card.[55] Emerging in the seventeenth century, these had become established by the eighteenth century as calling and business cards for tradesmen, also used as billheads, notepaper, and labels for instrument cases. Rococo decoration and incorporation of elegant figures was the key aesthetic for many such cards, showing how instrument makers invested in the imagery of polite science.[56] John Bird's trade card is a typical example, showing a beautifully illustrated quadrant within an ornate rococo frame (Figure 62). R. Rust's (fl.1745–76) card from the 1740s (Figure 63) goes a step further, showing an elegantly dressed gentleman taking a sight with an octant, reminding us of the figures used by both Barston and the *Universal Magazine*. A later card of Thomas Blunt (fl.1760–d.1823), in the 1790s (Figure 64), similarly shows a fashionable gentleman and lady inspecting his beautifully ordered shop in much the way that we can imagine Uffenbach did Hawksbee's decades earlier.[57]

Benjamin Martin's card (Figure 65), also from the 1740s, embodies best the interplay of instrument and image in creating these polite settings.

55 Fox has discussed these amongst other images of science in *Arts of Industry*, pp.239–42.

56 Snodin, *Rococo*, p.30.

57 Some trade cards were even useable, including calendars or wind roses. Walters, 'Tools of Enlightenment', pp.120–1.

Figure 65: Benjamin Martin, Trade Card (1740s); etching, 115 x 150mm, © Trustees of the British Museum.

Martin was the archetypal proponent of polite science, an instrument maker who also lectured and published widely. In 1771, another foreign visitor to London, Jean Bernoulli (1667–1748), discussed Martin's shop as 'one of the best furnished, where one is eager to attend his lectures … and his beautiful instruments make these doubly interesting'.[58] Martin's trade card showed this perfectly. Within a rococo cartouche, evocative of a piece of Chippendale furniture, a fashionable gentleman and lady strike elegant attitudes while using Martin's own scientific instruments to observe a star. Above them a detailed diagram shows the movement of stars, which we can imagine they are about to study. On the top right-hand corner of their cartouche stands a globe, of which Martin's 1765 publication, *Institutions of Astronomical Calculations*, might well have treated. Here Martin argued that

58 Bernoulli, *Lettres Astronomiques*, pp.72–3. Not everyone was so appreciative of Martin. Lalande reported that 'Mr Lespinass … told us about Mr Martin the optician, who gives ridiculous courses on physics', *Diary*, p.20.

> his *Celestial* GLOBE of twenty-eight inches Diameter is capable of such a Construction as must render it extremely useful at Sea ... Suppose he were to go one step farther, and take upon him to bestow upon this Globe that fashionable *Epithet* of the LONGITUDE GLOBE, he does not know that he should stand charged by Posterity, *one Hundred Years* hence, with greater *Absurdity, Vanity,* and *Presumption,* than the Authors of the LONGITUDE SECTOR, the LONGITUDE CHAIR, or the LONGITUDE CLOCKS.[59]

He shows how the mixing of instruments, images, and texts to perform polite interaction was also eminently part of the world of hack publishing and vitriolic criticism that we saw in Chapter 3. He criticises existing instruments and how they were perceived, clearly referencing Hadley, Irwin, and Harrison.[60]

CONVERSATION

POLITE DISCUSSION between men and women has already been noted as a hallmark of the public sphere. In the realm of natural philosophy, this was especially popular in the form of dialogues in which a male character instructed a female friend or relative. Starting with Bernard de Fontenelle's (1657–1757) *Entretiens sur la Pluralité des Mondes* in 1687, the genre was developed especially by proponents of the Newtonian philosophy and continued in works like Francesco Algarotti's (1712–64) *Il Newtonianismo per le Dame* in 1737.[61] This dialogue genre was central to establishing polite interaction between men and women in the discussion of natural philosophy, and marking out the roles that each should play in the conversation.[62] In his *Astronomical Dialogues* in 1729, derived from Fontenelle, John Harris made explicit the way in which the instruments that we have discussed so far, and which featured in his *Lexicon Technicum* that we encountered in Chapter 1, acted as the impetus for such discussions.[63] It is, in fact, the globes that

59 Martin, *Institutions*, pp.3–4.

60 The date of this publication is particularly significant, as Martin satirises 'Longitude' instruments in the very year that the Act was passed to recognise Harrison's and Mayer's achievements, with solutions that required the clock and sextant, and that Irwin's chair was discredited.

61 A 1728 publication of Fontenelle from The Hague featured a frontispiece with a polite lady and gentleman observing Cartesian vortices in the night sky.

62 Fara, 'Elizabeth Tollet', pp.169–87; Walters, 'Conversation Pieces', pp.121–54.

63 In 'Tools of Enlightenment', Walters makes clear how this included the astronomical prints by Senex, Martin, et al. The relative lack of explanation on these suggests they were intended to complement polite astronomical literature, p.256.

prompt the female character to ask for Harris' instruction, saying, 'they appear to be furnished up with that Curiosity and Care, that sure some very useful Knowledge is to be learnt from them, and is it not barbarous in you Men to confine it all to your selves?'[64] By using the term 'barbarous' she emphasised the role that women could play in 'civilising' hardened philosophers, while, in his introduction, Harris explained how men could make their learning attractive to the ladies: 'The *Digressions*, *Reflexions*, *Poetry* and *Turns of Wit*, are introduced to render *Those Notions* pleasing and agreeable, which perhaps without such a kind of Dress, would appear too crabbed and abstracted'.[65]

Benjamin Martin participated actively in this conversational side of polite science, bringing in the longitude problem directly. From 1754 to 1763, he published his 'The Young Gentleman and Lady's Philosophy' in which a brother, Cleonicus, instructs his sister. Her name, Euphrosyne, immediately invokes her namesake among the three graces of mythology, bringing associations of charm and merriment. The 'Philosophy' formed the first part of Martin's *General Magazine*, a monthly periodical made up of six individual half-sheets, intended each eventually to make up a separate re-bound collection. Its aim was to make natural philosophy readable in small, sociable portions.[66] The frontispiece to the *Magazine* showed brother and sister conversing in an elegant space, with the aid of the archetypal polite scientific instruments: a globe and telescope (Figure 66). In a series of dialogues, Cleonicus used such instruments to instruct Euphrosyne about astronomy, pneumatics, optics, and anatomy. In explaining the globes, Cleonicus detailed what the longitude and latitude lines meant, and Euphrosyne responded, 'I would not (though a Woman) speak improperly in naming this, or any other Part of the Sphere, or Globes'. She went on to ask, 'I have heard much Talk of the *Longitude*, and have as little Knowledge of it, after all, as though I had never heard it mentioned: But as we are now come to the Circle of Longitude, I presume you can give me a better Idea of that Matter'.[67] She made the much-discussed longitude problem a subject about which, as a polite woman, she should be capable of conversing with ease and understanding. Later, Cleonicus turns to the orrery to explain the planetary systems, and Euphrosyne asks, 'these *Moons of Jupiter* have been rendered very Famous on Account of finding the Longitude by them; now I should be glad

64 Harris, *Astronomical Dialogues*, p.5.

65 Harris, *Astronomical Dialogues*, p.v.

66 Millburn, *Benjamin Martin*, pp.68–71. 'The Young Gentleman and Lady's Philosophy' was re-assembled into two volumes and published in 1764.

67 Martin, 'The Young Gentleman and Lady's Philosophy' in the *General Magazine*, I, pp.184–5.

Figure 66: Frontispiece to Benjamin Martin, *Young Gentleman and Lady's Philosophy, in a Continued Survey of the Works of Nature and Art; by way of a Dialogue* (London, 1772); engraving, 213 x 123mm, Whipple Library, University of Cambridge.

to understand the Reason of that Affair. Pray, *Cleonicus*, could I be assisted in that Respect by this piece of Machinery?' Cleonicus' reply is telling: 'you may very easily; by Means of the Eclipses, which these *Moons* suffer in the Shadow of *Jupiter*; and to make the Thing as clear to you as possible, I shall place the Lamp in the Room of the Sun in the Centre to cast a Light on the Body of *Jupiter* and his little System of *Moons*'.[68] Cleonicus employed precisely the sorts of conversation, imagery, and instrument that we have been discussing, to explain the problem of longitude to Euphrosyne.

The dialogue form also impacted directly on longitude proposers. In 1720, Leonard Sturm wrote his *Project de la Resolution du Fameux Problème Touchant La Longitude sur Mer*, intending to submit it for the British prize as well as to the French Académie des Sciences. He constructed this in the form of a polite dialogue between a gentleman and a mathematician, in which the two discuss 'Sturm's' project. He explained in the preface that he had chosen the dialogue as an easy method of making the truth clear.[69] 'Sturm' is supposedly known to both the gentleman and mathematician and so, by distancing himself as author from himself as character within the text, Sturm constructed his proposal completely through the discussion of the two protagonists. He framed the dialogue with the story that the Gentleman had been ridiculed at Court for not understanding discussions of Sturm's project and had gone to the mathematician for instruction. The gentleman explains, 'We were sitting at table, where I was preparing the food, and poor Sturm was given a hard time because of the Longitude, which the others said he claimed to have found at Sea, although he has got no acceptance for his pains' ['bien qu'il s'en eut attiré bien de la prostitution'].[70] (We note that the French uses the idea of 'prostitution' to discuss Sturm being socially rejected for what are seen as improper theories, an association of ideas to which we will return in Chapter 6.) The gentleman continues to visit the mathematician for a series of conversations in which they discuss the shape

68 Martin, 'The Young Gentleman and Lady's Philosophy', p.206. This is also reminiscent of Joseph Wright of Derby's painting completed around the same time, *A Philosopher Giving That Lecture on the Orrery in Which a Lamp Is Put in Place of the Sun* (c.1766); oil on canvas, 147.3 x 203.2cm, Derby Museums and Art Gallery. Walters has discussed in 'Tools of Enlightenment' how the original representation of the orrery by Thomas Wright was continued by other makers, who also recommended dialogues as instructive materials, pp.140–3, 229–36.

69 Sturm, *Project*, Preface p.2.

70 The full French reads, 'Nous étions à table, ou je coupai les viandes, & alors le pauvre Sturm passa par le Serain à cause de la Longitude, qu'il prétend, disoient ils, avoir trouver sur la Mer, bien qu'il s'en eut attiré bien de la prostitution', Sturm, *Project*, p.1. My thanks to Adrienne Hamy for help with this translation.

of the earth and the construction of longitude and latitude lines, as well as the relative success of different methods to find longitude, and of specific English and French proposals. Those of Whiston and Ditton,[71] and of Alimari,[72] are criticised specifically. The gentleman ends by recommending that Sturm keep his invention private until he can find a patron, while the mathematician maintains that his knowledge should be open to public discussion. He advocates, in fact, precisely the polite conversation in which he and the gentleman are involved.

It is unusual, however, to find such dialogues between two men for, as we have seen, one role of polite conversation was to encourage controlled and educational interaction between the sexes. Yet the underlying question remained to what extent women had the mental capacity and moral fibre to participate in these conversations; how much, in fact, they could and should know.[73] Two works published in the 1720s and '30s in support of women's intellectual equality with men certainly used scientific knowledge and methods as a means of asserting this. In *Woman Superior to Man*, published in 1745, the author argued for distinctively Lockean scientific reason as necessary to consider difference between the sexes (we must bear in mind that the author's female identity may well be a persona, a practice so often beloved of eighteenth-century writers):

> to set this whole matter ... in as clear light as possible, it will be necessary to clear our ideas from all that is huddled and confused, by separating the fictitious from the real, the obscure from the evident, the false from the true, supposition from matter of fact, seemings from entities, practice from principle.[74]

The pamphlet further argued that women had particular skill in communicating science, as 'to a taste for the sciences they join such a happiness in the use of them, whether conversing or acting, as is sufficient to excite others to the study of them'.[75] The earlier *Woman Triumphant*, published in 1721, had likewise used natural philosophical knowledge as a key example of a well-brought-up young woman, pointing out 'how many Examples have we in the *Ladies Diary*, of Gentlewomen who are perfect *Oedipus*'s at unfolding Riddles, and very *Sphinx*'s for inventing them?'[76] Both therefore made

71 Sturm, *Project*, pp.25–6; he references Whiston's *New method*.
72 Sturm, *Project*, p.13; he references Alimari's *New method*.
73 Fara, 'Elizabeth Tollet'.
74 'Woman Superior to Man' in *Beauty's triumph*, p.9.
75 'Woman's Superior Excellence over Man' in *Beauty's triumph*, p.24.
76 Lady of Quality, *Woman triumphant*, p.15.

Figure 67: Plate I from William Hogarth, *The Analysis of Beauty* (1753); engraving, 390 x 505mm, gift of Sarah Lazarus, 1891, Metropolitan Museum of Art, New York.

female knowledge of natural philosophy both an attribute and a necessity in performing politely.

Hogarth also exhibited such ideas of polite interaction in his book *The Analysis of Beauty*, published in 1753 to argue for his own theory of aesthetics. In *The General Advertiser* of November 1752, he explained that 'he had endeavoured to render it useful and interesting to the Curious and Polite of both Sexes, by laying down the Principles of personal Beauty and Deportment, as also of Taste in general, in the plainest, most familiar, and entertaining Manner'.[77] The two engraved plates for the *Analysis* placed men and women in precisely the polite settings we have discussed, conversing about art in a sculpture yard (Figure 67), and dancing in a grand picture gallery (Figure 68). Both surrounded these images with instructive text and illustration, much like the *Universal Magazine*, as well as the pamphlet

77 Quoted in *Analysis*, p.xvii.

Figure 68: Plate II from William Hogarth, *The Analysis of Beauty* (1753); engraving, 425 x 535mm, gift of Sarah Lazarus, 1891, Metropolitan Museum of Art, New York.

illustrations that we considered in Chapter 2. The *Analysis* also dealt specifically with women. The 'line of beauty' was particularly drawn from the human body rather than the classical statue or mechanical object, and especially the female body rather than the male.[78] Hogarth defined this line with the corseted female form, giving a spectrum of examples (Figure 67).[79] He also compared female skill in observing and displaying this line directly to scientific observation, saying, 'the ladies always speak skilfully of necks, hands, and arms; and often will point out such particular beauties or defects in their make, as might easily escape the observation of a man of science'.[80] Yet attention to the human body strayed dangerously into the impolite behaviour in society.

78 Snodin has also pointed out how rococo the line of beauty is, see *Rococo*, p.32.

79 *Analysis*, pp.48–9.

80 *Analysis*, p.68.

An Impolite Science

In 1738, James Miller (1706–44) published his poem *Of Politeness*, in which he advised on how a young man could perform this attribute. The longitude featured as one of his key examples of impoliteness, in the example of a man too obsessed with learning, who

> Wond'rous deep Schemes, like *Whiston*, can impart,
> And bring to Light the Myst'ries of each Art:
> Reveal the Longitude, the Circle square,
> And [make] … the Fair
> Know all Sir *Isaac Newton* to a Hair.[81]

In this case, it is the young man's mode of conversation with women and his wish to 'reveal the longitude' that create an impolite situation. This brings us to the flip side of the polite conversations, imagery, and instruments of science, and the space where it seemed longitude could play a key role in more ribald, misogynist, and even erotic literature, as a means of expressing fears around female activity in both science and society. In Bedlam, our two women coyly ogle the semi-naked form of the longitude lunatic. They are not just passively absorbing the events around them but actively participating. The cheap copies of the prints that Thomas Bakewell produced on Hogarth's instruction pointed this out specifically, saying, 'one is whispering the other to look at the naked Man'.[82] Politeness required repressing the realities of the human body, but here we have the women as voyeurs.[83] Their active sexual presence is linked to their active scientific interest, and it is longitude that is the focus of their attention.[84]

Both natural philosophical spaces and instruments were already important to establishing such boundaries between the polite and the impolite in the eighteenth century. The seventeenth-century republic of letters was formed through the social codes established in dinners, letters, sharing of objects and information, but also in the 'delighted disgust' with which the breaking of social codes was viewed.[85] It was just one of a number of sites and institutions that provided a 'thinking space' for politeness to be articulated and delineated.[86]

81 Miller, *Of Politeness*, p.6.
82 Bakewell, after Hogarth, *A Rake's Progress* (1735), colophon to Plate VIII.
83 Walters, 'Conversation Pieces', p.136. These women can also be compared to the Magdalene figure of Sarah Young, weeping over Tom Rakewell. See Steintrager, 'Perfectly Inhuman', pp.123–4.
84 Price, 'Patterns of Sexual Behaviour' in Boucé (ed.), *Sexuality in Eighteenth-Century Britain*, p.170.
85 Goldgar, 'Impolite Learning', p.9.
86 Stallybrass and White, *Politics and Poetics*, p.80.

The kind of words with which politeness was opposed by contemporaries—'vulgarity', 'rusticity', 'barbarity', 'productive', 'useful'[87]—become particularly problematic in the context of natural philosophy, especially when those performances were reliant on the presentation and control of a human body, itself an instrument of science.[88] Such words also show how much questions of politeness tie into the kinds of ephemeral and satirical literature, of projecting and hack writing that, precisely, existed on the margins of the polite world, and which we have seen as integral to discussions of longitude. These works not only questioned the status of the longitude problem as knowledge, but also how it could be sought after, experimented on, and presented within the constraints of polite society, or indeed whether it should be. Erotica was a particularly expressive form to articulate such questions, showing the human body, and especially the female body, both straightforwardly and satirically as a site in need of particular attention and control. This is also a site where natural philosophy, and longitude, appear surprisingly frequently, as erotica became a means to attack innovation.[89] Politeness depended on the establishment and discussion of boundaries, the measurement of what was appropriate in which setting, and the exhibition only of what was agreed to. Longitude, and latitude, became apt metaphors for articulating such conflicts.

'Why is a Woman like a Mathematician?'

We have considered female participation in conversation as intrinsic to polite natural philosophy, meant to hold a civilising influence over the rough and pedantic male, while playing the part of the informed but submissive companion. Intrinsic to this idea, however, was the understanding that women's active participation in science could go only so far.[90] As M. Vaumorière (1610–93) had commented in his 1691 publication, republished

87 Klein, 'Politeness for Plebes' in Bermingham and Brewer (eds.), *Consumption of Culture*, pp.365–6.

88 Delbourgo, *Most Amazing Scene*, pp.87–127, quotation pp.119–20; Stafford, *Artful Science*, pp.133–215.

89 Wagner, *Eros Revived*, pp.191–9; Peakman, *Mighty Lewd Books*, pp.67–92; Dunn, *The Telescope*, pp.114–19, and 'Taming the Telescope', pp.191–201.

90 Schiebinger has discussed astronomy as one of the 'male' fields of knowledge in which exceptional women could excel, but generally only under the patronage of a male relative, which they could not then take on to any public role (unlike some black men), see *Nature's Body*, pp.191–2. This compares to the analysis that Rogers has made of women playing male parts in eighteenth-century theatre, where donning male habits only became sufficiently convincing to enforce gender divides. See 'The Breeches Part' in Boucé (ed.), *Sexuality in Eighteenth-Century Britain*, pp.244–58.

in 1722 and 1735, that advised women on *The Art of Pleasing in Conversation*, 'I am of opinion that it is very consistent with their modesty to be perfect masters of … Geography, provided they don't aim at appearing too much Geographers, by affecting hard words too rugged for their fine mouths … I should not like to have them frighten me with *Longitude* and *Latitude*'.[91] In 1739, 'a Gentleman' published *Man Superior to Woman* in direct reply to the work *Woman Superior to Man* discussed above. He laid out in no uncertain terms that he thought women incapable of studying science in any meaningful way, saying,

> a severer Imprecation cou'd scarce be utter'd against the lovely Sex than to wish them Science-mad. 'Tis their pretty fluency in Nonsense, and their bewitching Confidence in Ignorance, which give their Charms the Power of pleasing us … Why shou'd they be suffer'd to distract their poor tender Brains with hard Words and technical Terms.[92]

Note that his sole concern here was which female attributes would be pleasing to a man. Furthermore, he used longitude and latitude as the tools to belittle female interest in science. He described theory as

> a Ground too holy for Female Feet to tread with Impunity … therefore I shall wholly confine myself, in this little Treatise, to practical Reasoning, except where I am obliged to step aside to recover my bewildered fair Antagonist from the Danger of straying out of her Latitude,[93]

and he went on with dripping sarcasm to discuss how

> you will find … [women with] the very Dregs of our Sex; Fops whose whole Merit is made up of Dress and Drivel, Shew and Emptiness … eminently qualified to … square the Round of a *Woman's* Petticoat, and take the Latitude of a Nightcap from the Equinox of her Noddle, or the Longitude of two Lappets by the Meridian of her Whims.[94]

Ideas of longitude and latitude are key to the exaggerated language used to articulate female behaviour seen as outside polite territory. This compares beautifully with Matthew Darly's (1720–78) 1771 print, in which a lady's hair arrangement is so absurdly tall that her male companion is forced to use an octant, inexpertly, to admire it (Figure 69).

A wealth of less vitriolic sources in the period similarly satirised female understanding and interests as comparable to the ridiculous search for the

91 Vaumorière, *Art of pleasing*, ii, pp.131–3.
92 Gentleman, *Man superior to woman*, p.46.
93 Gentleman, *Man superior to woman*, p.7.
94 Lady of Quality, *Woman triumphant*, pp.30–1.

Figure 69: Matthew Darly, *The Ridiculous Taste or the Ladies Absurdity* (1771); etching, 370 x 260mm, courtesy of the Lewis Walpole Library, Yale University.

longitude. A selection of examples serves to show the sheer number and variety. In 1759, in *The Life and Real Adventures of Hamilton Murray*, the author replied to an elderly lady on the catch for a husband, 'Why Madam, to be sure as you say, you are of the fashion, but what kind of fashion I believe will be as puzzlesome to find out as the longitude'.[95] Conversely, in Elizabeth Cooper's (fl.1737) *The Rival Widows*, Freelove commented to Lady Bellair, 'Your Ladyship will allow, I had a shrewd Guess to find out a Woman's Mind' to which she replies, 'Ay, if you had found it out; but I'm of Opinion you'll discover the Longitude as soon'.[96] Women's capacity for constant, and inane, chatter was also compared to perpetuum mobiles, and therefore to longitude. In 1742, *The Universal Masquerade* joked, 'consider how much Time and Pains

95 Murray, *Life*, ii, p.46.

96 Cooper, *Rival Widows*, p.73.

have been thrown away upon a North-east Passage, the Longitude, and the Perpetual Motion. Tho' some modern Philosophers, who are marry'd Men, are full of Opinion, that this last is already discovered'.[97] In *The Curiosities of Paris* in 1760, the author described French women as 'sprightly and witty in Conversation … but a little too talkative. This hint may perhaps be useful to the mathematical Gentlemen who are now labouring to find out the Longitude; for almost every pretty French Lady can furnish them with a *perpetuum Mobile*'.[98] In 1768, another author listed in *An Original Camera Obscura* 'several poetical Essays [including] … On Women's *Tongues*: Proving to a Demonstration, that they are the best Machines by far to discover *the Longitude*. Intended as an useful Hint to the *two Houses* to save the Nation's Money, as *Electioneering* is at hand'.[99]

Furthermore, it was a woman's fickleness in love that was constantly linked to the difficulty of finding longitude: as difficult to find as her heart. In the play *The Fashionable Lady*, Trifle rallied his friend Mr. Whim on being rejected, saying, 'verily you look as mortified, as a discarded Member of the Royal Society, or a Mathematician disappointed of the Longitude'. To which Whim replied, 'What a Fool was I to dream of Comfort in Matrimony?'[100] In *The Third Volume of the Works of Mr. Thomas Brown*, the author opined, 'What could my grave considering Friend John, at past five and forty, place his happiness in a Woman? … Is the inconstancy of the Sex, after all, so little known to him, whose Mind veers every Moment, runs round the Compass, so soon as the Sun round the Earth? You might as easily fix the Longitude, as a Woman's mind'.[101] James Miller took this a step further in his play *The Humours of Oxford*. Here, Gainlove tells his friend Trumore at the start of the play that, ''tis easier to discover the Longitude, than the Situation of a Woman's Heart – and their Inclinations vary ten times oftner than the Weather-glass',[102] but then at the *denouement* finds himself tested as to his fitness as a husband based on that very knowledge when his potential mother-in-law asks, 'but one thing more … and you shall be conducted to your bride … Do you think it ever possible to find out the Longitude, Sir? – it is such a vast Profundity, that I fear 'tis beyond the reach of any Man to fathom it'.[103] Lastly, John Stevens' (fl.1744) play *The Modern Wife; or, The Virgin her own Rival*, opened with 'Friendly' bemoaning

97 Sedgewick, *Universal Masquerade*, ii, p.14.

98 A. R., *Curiosities of Paris*, pp.146–7.

99 R*. R*., *Original camera obscura*, p.33. This joke was copied almost exactly in a footnote to Martial, *Epigrams*, pp.172–3.

100 Ralph, *Fashionable Lady*, p.79.

101 Brown, *Third Volume*, pp.79–80.

102 Miller, *Humours of Oxford*, p.2.

103 Miller, *Humours of Oxford*, p.58.

of his friend Sir George Modern, 'Ah! Cursed Matrimony! What a deplorable Metamorphosis hast thou wrought? The once gay, merry, facetious Modern, is all on a sudden transformed into a dull, melancholy Dotard—I know his Meditations :--They are all on Women; and he might as well study the Longitude, or the Philosopher's Stone'.[104] Note the sexual element in the title here, to which we will return.

It is, therefore, unsurprising to find not only that longitude featured in two popular jokes from *Ben Johnson's Jests* published in 1751, but also that both of these combined this humour with a jibe against women.[105] The first related how

> Some Pretenders to the Mathematicks, at a Tavern near St. Paul's, were debating t'other Day about the Longitude … and one of these mighty Philosophers observ'd, that the Longitude might be easily discover'd, cou'd they but find the perpetual Motion; upon this the Drawer ran out, and calling his Mistress, Here, Gentlemen, says he, here is the perpetual Motion for you, and now I hope you'll find the Longitude. Madam was angry, and taking up the Bottle, knock'd him down, There, Sirrah, says she, learn to know your own Latitude, and for the future don't make so free with your Superiors.[106]

The second joke asks 'Why is a Woman like a Mathematician?'[107] and the answer comes, 'because she endeavours to find out the Longitude'.[108] Both jokes have further connotations. The drawer's mistress not only takes charge over her husband, physically and socially, by knocking him down and reminding him of his manners before his superiors, but she also uses latitude as a means of reminding him where these boundaries lay. Longitude might be used as a means of highlighting or satirising women's mental limits and foibles, but female commentators also consciously used longitude as a means of asserting their power and mental abilities. In 1756, in *The Husband*, Eliza Haywood (c.1693–1756) turned male obsession with female interest in fashion back onto her male critics, and used longitude as her parting shot, saying,

> there are some people who, to acquire the reputation of a man of taste … waste three parts in four of their time in consulting with their taylor … and if they hit on any invention which happens to take the fancy of the town … value themselves more upon it than if they had found out the Longitude.[109]

---

104 Stevens, *Modern wife*, p.1.

105 Wagner has written about the wider genre of such jokes in *Eros Revived*, pp.167–71.

106 *Ben Johnson's Jests*, p.50.

107 *Ben Johnson's Jests*, p.80.

108 *Ben Johnson's Jests*, p.105.

109 Haywood, *Husband*, pp.184–5.

We are reminded of Uffenbach's censures on Woodward at the beginning of the chapter. Similarly, in *Woman's Superior Excellence over Man*, 'Sophia's' further reply to the author of *Man Superior to Woman*, she equated male rejection of female knowledge and skill to bad scientific practice, arguing that 'one grand vulgar error which has crept into *Philosophy*, thro' the extravagance of mannish learning, is the very great distinction made among the sciences … If we hunt this chimera up to its source, we shall find it to proceed from the same cause with that which influences their weak opinion of *Woman-kind*, to wit, *Prejudice*'.[110] We might conclude, then, that such writers could no more understand a woman's mind than the longitude because, in fact, their 'mannish learning' had made them reject both prizes as impolite.

'THE SO MUCH DESIRED LONGITUDE'

THE WOMAN in the second of *Ben Johnson's Jests* who 'endeavours to find out the longitude' clearly seeks a more physical transgression, and the gendered interactions of polite science could, of course, also be the route to an intrigue.[111] John Harris' female companion in the *Astronomical Dialogues*, indeed, openly flirted with him, saying, 'you are really, Sir, a very bad *Woman's-Man*; you have Philosophised me out of many a fair Pleasure already; Censure, Satyr and Gossiping are almost gone; and must dear *Inquisitiveness* follow them too?'[112] She made clear their mutual understanding that Harris and his male compatriots enjoyed the illicit nature of such flirtations, at the same time as marking out the female characteristics understood to be impolite. Such flirtations also entered the sorts of scientific publications within which we have already encountered the longitude problem. Robert Heath's vitriolic *British Palladium*, which we considered in Chapter 4, began life as part of the *Ladies Diary*, the annual publication of mathematical problems that is eulogised above by the author of *Woman Triumphant*, and which is used by many historians as a hallmark of female polite science. In 1752, Heath was removed from his position as editor after allowing the publication of a problem in particularly questionable taste, if not writing it himself.[113] In that year, Question 359 ran thus:

> Miss's Apron grown short, she is full of Complaint,
> And to merit your Pity she looks like a *Saint*! …

110 'Woman's Superior Excellence', p.77.
111 Secord, 'How Scientific Conversation became Shop Talk', p.144.
112 Harris, *Astronomical Dialogues*, p.31.
113 Ruth Wallis, 'Robert Heath' in *ODNB*.

Mamma now was told *Miss*'s Hoop was too small,
Therein lay her Grievance, Disorder, and all …
A Hoop of the Fashion to cure her Disease,
Extends from her *Centre* quite round to the Knees: …
Her hidden Perfections she knows will invite,
Or ensnare the Beholder, should Chance give them sight.
By the Pow'r of her Hoop *Polly* steps into Fame,
By out-priding the rest *she conceals her own Shame* …
Her *Hoop* is the Secret – and if you would know
What it holds with her *Petticoat*, seek from below.[114]

The problem posed, after all this discussion of a young woman's skirts and what was shown or hidden underneath them, was simply to calculate the volume of the petticoat.[115] Heath articulated precisely the concealing and performative nature of politeness that we have been discussing.[116] It is no wonder that Ferrner reported the mathematician John Robertson telling him that Heath 'was not a bad mathematicus; but of such a bad character and disposition that he did not deserve anyone's company'.[117]

The faintly erotic elements of this satire bring the longitude problem into the 'sexual underworlds' of the enlightenment. It is just these kind of pamphlet materials that reveal for us sexual ideas in texts beyond the standard literary canon, and that particularly used scientific discussions as a veil for pornographic interest.[118] If we look again at the pirate versions of *A Rake's Progress* made in 1735, we see that Hogarth's copiers remembered our Bedlam women performing much less nuanced roles than those Hogarth had given them (Figure 3).[119] The pirate Plate 8 shows one old, withered woman by herself who is using a mirror simultaneously to admire her own face and to look at the man in the cell. This plate's author remembered the woman solely as a voyeur interested in her own body and that of the men around her. It is also noticeable what the man in the cell is doing in this

114 *Ladies Diary*, pp.25–6.

115 The petticoat was a subject of contemporary satire, seen as emblematic of the ridiculous heights (or widths) which feminine vanity had reached. See for example A. W., *Enormous abomination*. This is discussed by Wagner in *Eros Revived*, pp.157–8.

116 *Ladies Diary*, thus, may have been a polite literary space, but it was also a vehicle for gentle or ribald satire of public mathematics and its practitioners in the established genre of almanacs such as the long-running *Poor Robin* produced from 1665. Benjamin Wardhaugh has discussed this relationship, and linked both publications to the *Nautical Almanac*, in *Poor Robin's Prophecies*, pp.132–5, 169–75.

117 Orrje, 'Navigating the Neighbourhoods', pp.27–30, 40. Walters discusses a similar pamphlet attributed to Heath in 'Tools of Enlightenment'.

118 Rousseau and Porter, *Sexual Underworlds*, p.14.

119 The swooning Sarah Young is also a much less modest figure in the piracy.

version. There is no stream of urine as in Hogarth's plate, but instead he has his hands unambiguously down his trousers. Indeed, it has been argued that Hogarth made such a change himself in the 1763 re-working of the plate.[120] The commentator George Steevens (1736–1800) certainly reacted with disgust to Hogarth's image in 1781 in his *Biographical Anecdotes of William Hogarth*, where he asked, '[H]ow is the hand of the ideal potentate employed while he is gazing with no very modest aspect on a couple of young women? And to what particular object are the eyes of the said females supposed to be directed?'[121] So, to some contemporaries these women were watching a man masturbate, making them voyeurs of a sexually knowledgeable and active kind.[122]

This shows the sexual awareness encoded, and hidden, in flirtatious polite intercourse, where female sexuality was acknowledged by making it transgressive. Women were not merely the sexual, scientific objects displayed in John Lodge's (d.1796) 1769 engraving of *A Philosopher giving Lectures on the use of the Globes*, in which the woman's breasts are the 'globes' that the male company observes with so much interest.[123] In 1746, a small pocket leaflet was published entitled *Wit a la mode*, offering a series of humorous questions for ladies, along the same lines as the jokes in *Ben Johnson's Jests*. It was easily carried in the pocket and disguised as cards. A second leaflet offered the answers (Figure 70). All the questions cast aspersions on female chastity, such as 'Why is it impossible to ravish some Women? ... Because they are willing'. Another asks, 'Why are women fitter for the Study of Astronomy than Men?' and the answer comes 'Because they lie more upon their backs'. Notably, the questions are supposedly sent by one woman to another as amusements given by a male acquaintance. Thus, not only is women's scientific knowledge associated with easy virtue, but women are shown empowering each other in the pursuit of this. Earlier we met Eliza Haywood using longitude as a boundary to critique male interests in *The Husband*. She both actively encouraged communal seeking of knowledge by women without the aid of men and wrote a series of erotic novels.[124] Thus, there were spaces, both

120 Stevenson, 'Hogarth's Mad King', pp.24–43.

121 Nichols, *Biographical Anecdotes*, p.38.

122 This suggestion has also been supported by Krysmanski in *Hogarth's Hidden Parts*, pp.167–70.

123 John Lodge, *A Philosopher giving Lectures on the use of the Globes* (1769); etching, 210 x 130mm, Lewis Walpole Library.

124 Schiebinger, *The Mind has no Sex?*, p.32; Peakman, *Mighty Lewd Books*, pp.36–7. See also Koon, 'Eliza Haywood', pp.43–55. Haywood likewise wrote romance novels, which Patricia Meyer Spacks has described as 'semi-pornographic' in 'Ev'ry Woman is at Heart a Rake', pp.32–3.

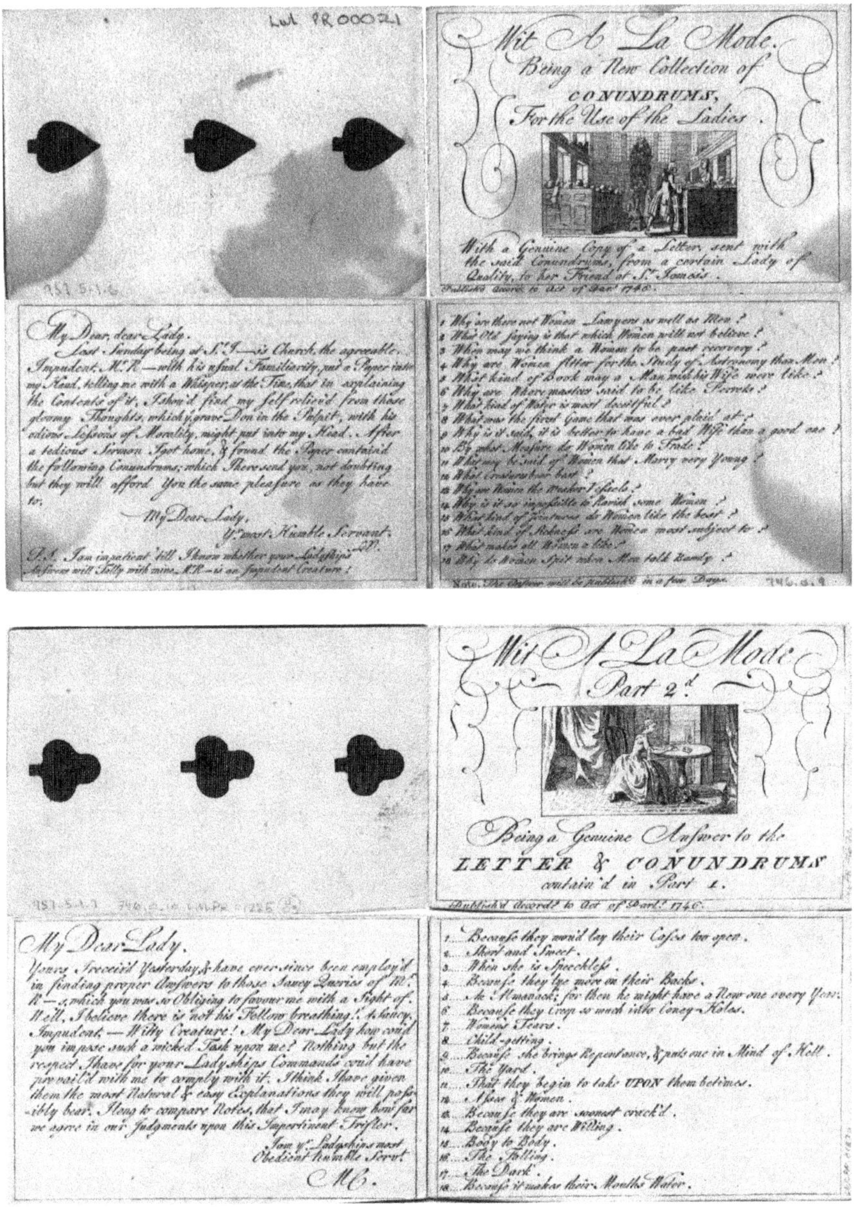

Figure 70: Anon., *Wit a la mode: being a new collection of conundrums, for the use of the ladies with a genuine copy of a letter sent with the said conundrums, from a certain Lady of Quality to her friend at St. James's* (1746) and *Wit a la mode: being a genuine answer to the letter & conundrums contain'd in Part 1* (1746); front and back; engravings, both 70 x 200mm, courtesy of the Lewis Walpole Library, Yale University.

physical and literary, in which women could be the bearers of knowledge, but in those spaces they would also be feminised as a sexual object.

In 1718, John Breval's (c.1680–1738) piece, *The Play is the Plot*, featured a young man, Peter Pirate, intriguing to marry Prudentia, a rich widow. Pirate does this through pretending to share Prudentia's interest in natural philosophy and linking this to their mutual desire. He exclaims,

> I hope … you will pity a Man that's imprison'd in so much Body, and urg'd by such violent Passions; I am drawn to your Ladyship, I must own, by a kind of Centripetal force, and will communicate a Secret to you, which none but your self have extorted out of these Lips—I have found out the Longitude … ready Money to try some Experiments upon *Hampstead Heath* now is the only thing wanting, in order to lay it before a Committee.

To which Prudentia rhapsodises, 'The Longitude! I am transported at the thought of it, Sir! … but to follow me into my Closet … and we shall be able to discuss an affair of that consequence without Interruption'.[125] Not only does the longitude allow Peter to announce his desire for her, 'imprison'd in so much Body', but it also gains him access to her closet. He achieves this through reference to precisely the kind of public experimental displays that we discussed above as practised by William Whiston. The fact that Peter is a pirate copier of engraved prints makes him especially dangerous when let into the closet of a woman's mind, as we saw in Chapter 2.[126]

Indeed, a number of works compared finding longitude to the difficulty of persuading a woman to let a man into her 'closet'. In *Jemmy Carson's Collections*, published in 1744, the author included a supposed letter advising an elderly friend against remarrying a young wife. He presented consummating such a marriage explicitly as an act of violation, imagining how,

> At thy first going to Bed, thee may probably, through Curiosity … find the Philosophers Stone at each side of the Gates of *Babylon* – Paradventure, she will say unto thee, struggling, *Oh!* Obediah! *they must not part* … I will now intreat thee in *Philosophical Terms*, and desire thee mayest suppress this *Topsy Turvey* in thy *Members*, that threaten so much Destruction … for the latter is a mere *Chaos*, where *Longitude* is without *Perpetual Motion*, which the Infirmities of old Age will not allow thee to maintaine … 'tis easy for thee to take the *Altitude of the Sun* with an Instrument: But thee will never find an Instrument, of what Magnitude soever, to … find out [the] Diamiter of thy Wife's *earthly Semi-Circle*.[127]

---

125 Breval, *Play is the Plot*, p.22.

126 Mitchell has suggested that marriage 'copyrighted' women for sex in *Death by Hogarth*, p.52.

127 *Jemmy Carson's Collections*, pp.82–3.

Note how both the male and female body in this are equated with instruments, as well as with longitude and perpetual motion, and how both are part of a 'Topsy Turvey' transgression of social norms.

In the light of such discussions, it is less surprising to find a number of erotic poems from the eighteenth century which represent women using the longitude as a means of taking not only intellectual but also sexual power.[128] In a manner in which the author of *Man Superior to Woman* so feared, and for which Robert Heath seemed to hope, the process of these women doing science was intimately linked with exhibiting their sexuality. In 1726, *The Longitude Discover'd; A Tale* was published, a poem which started from the point of the 1714 Longitude Act, describing men from all walks of life leaving their professions to seek the great reward. The best hope of success belongs to 'Astrophel, the old and wise, / Whose knowledge reached beyond the skies',[129] who decides to marry on the verge of old age.

> E'er honey-moon's half spent, he flies,
> From love (that Feast which sometimes cloys),
> To find the Longitude, prepares,
> Old Age's Eve to free from Cares …
> This Project brought the hop'd Relief,
> Of Books and Maps he turn'd the Chief,
> But not one Matrimonial leaf.[130]

Thus, he pays little attention to his young wife who, dissatisfied, looks around for a younger man and finds one in Church (a further transgression). They become lovers and increasingly reckless in their meetings, such that she plots to bring him into her marital home. She tells her philosopher husband how worried she is that he works too hard and offers to make observations in his place. Meanwhile, her lover comes to the observing room. She thus makes a mockery of the model 'guild wife' that historians have identified acting as 'invisible assistant' to her philosopher or artisan husband.[131] The noise of the couple's lovemaking disturbs the husband, who calls,

> … What Noise?
> My Dear! – (she readily replies)
> Success, I hope, and very soon,
> The *Dragon's-Tail* is near the *Moon* …

128 Wagner has discussed the growth of such literature as a marker of male fear at increasing female sexual freedom, *Eros Revived*, p.143. Increasing freedom to express and celebrate sexuality is discussed by Dabhoiwala in *Origins of Sex*.

129 Author of The Deluge, *Longitude discover'd*, p.4.

130 Author of The Deluge, *Longitude discover'd*, pp.4–5.

131 Schiebinger, *The Mind has no Sex?*, pp.66–101, 260–4.

All these Conjunctions promise well,
(The Dotard cries:) My dear, pray tell,
How far from *Mars* is *Venus* off?
About nine Inches and a half,
(Replies the Dame:) Sleep but one Hour,
The Longitude I shall secure.[132]

We see the fairly obvious joke that can be made using longitude to compare her lover's prowess to her husband's, who has previously been discussed as having 'sapless Back, and weakned Reins, / His Vigour all lay in his Brains'.[133] She 'endeavours to find out the longitude' like the woman in *Ben Johnson's Jests*, with both meanings that entails. Then disaster strikes as the young lovers grow more abandoned:

His vigour gave his Love the Slip;
Down came the Table, Books and Globes,
The Telescope and Astrolobes;
With dreadful Noise, and shocking Sound,
The falling Planets strew the Ground ...
Scar'd *Astrophel* peeps in at Door,
And sees his Trump'ry on the Floor.[134]

We can here note that the 'Trump'ry' represents both Astrophel's instruments and his wife. She claims, to explain the chaos, that she had just found the longitude and that a sprite had come through the window and stolen it: 'So near the Brink of Luck to stand! The Longitude was in my Hand!'[135] The elderly fool cuckolded by a young wife is, of course, not a new comic trope in the eighteenth century,[136] but it is not simply that the philosopher is being ridiculed for his inability to interact with the real world embodied by his young wife. It is his search for longitude that is presented as emasculating him: the longitude he might have achieved sexually with his wife is replaced by a longitude of impossible dreams. The wife is, by contrast, empowered by her control of both the longitude and her own body. Her joke that she is 'nine inches and a half' from the longitude gives her control over her husband's acquisition of knowledge, both scientific and matrimonial. She turns on its head the contemporary equation of the virgin female body with a virgin mind, both innocent and ignorant.[137]

132 Author of The Deluge, *Longitude discover'd*, p.10.
133 Author of The Deluge, *Longitude discover'd*, p.5.
134 Author of The Deluge, *Longitude discover'd*, p.11.
135 Author of The Deluge, *Longitude discover'd*, pp.11–12.
136 Wagner, *Eros Revived*, pp.143–5.
137 Spacks, 'Ev'ry Woman is at Heart a Rake', pp.27, 40.

In 1739, J. Jones (fl.1739) published a considerably more explicit adaptation of this poem entitled *The Star-Gazer: or Latitude for Longitude*. This made the link between longitude and sex, if possible, even more explicit, discussing the young bride's motives for entering such a marriage:

> 'Twas duty made her give consent:
> Her Mammy thought it a great Honour
> To have this great Man lye upon her.
> And on a Chariot did conclude, –
> If he found out the LONGITUDE.[138]

This reminds us of the answers found by using cards in *Dr. Flamstead's and Mr. Patridge's New Fortune-Book*. Jones also gave his heroine a more active interest in both sex and natural philosophy, and again made this mutually dependent, when in justifying her offer to observe for her husband she says,

> Besides, I've got a mighty Notion
> Of that same Thing, Perpetual Motion,
> And often wish that I might find it,
> Tho' you may fancy I ne'er mind it;
> On it I'm thinking, tho' not speaking,
> For many a Night it keeps me waking.[139]

As in *Jemmy Carson's Collections* above, perpetual motion becomes not a joke about women's incessant chatter, but about a man's inability to perform sexually. Both make the same joke as the 1772 print *Doctor Blowbladder discovering the Perpetual Motion*, in which the young woman's vital body is compared to his aged lechery (Figure 71). Note the size and placing of the doctor's stick and the medicine bottle that falls out of his pocket.[140] With such multiple meanings associated with the perpetual motion, we might think here about the common eighteenth-century myth that the size of a woman's vagina equated to the size of her mouth.[141] Does the frequency of movement of one also equate to the other?

Yet comparison was also commonly made between women who had been overly sexually stimulated and hermaphrodites, making both monstrosities. Monsters portrayed not only physical but also mental deformity, and contemporary women who indulged their passions were, therefore, seen as insane and thus also inhuman (linking, of course, to discussions in the

138 *The Star-gazer*, p.4.

139 *The Star-gazer*, p.8.

140 This also links to discussions of medical projecting in Chapter 3.

141 Boucé, 'Some Sexual Beliefs' in Boucé (ed.), *Sexuality in Eighteenth-Century Britain*, p.33.

Figure 71: Anon., *Doctor Blowbladder discovering the Perpetual Motion* (1772); mezzotint, 373 x 252mm, Wellcome Collection, Public Domain.

previous chapter). Women with philosophical skills could, therefore, become less sexualised, like a hermaphrodite.[142] This relationship between female scientific knowledge and sexuality even emerges among Swift's projecting satire in *Gulliver's Travels*.[143] Deborah Armintor (fl.2004) has argued that the microscopic nature of Gulliver in Brobdingnag makes him analogous to the small pocket microscopes fashionable among women who were interested in polite science in the eighteenth century. This she links to the degradation of

142 Masturbation was supposed to enlarge the clitoris and therefore make such women appear like hermaphrodites. See Wagner, 'Veil of medicine and morality', pp.181–2. Spacks discusses sexually indulgent women as insane, in 'Ev'ry Woman is at Heart a Rake', p.37. Christina of Sweden was called a hermaphrodite because of her philosophical skills, according to Schiebinger in *The Mind has no Sex?*, p.47.

143 Armintor, 'Sexual Politics', pp.619–40.

the microscope as a tool for 'serious' science, contending that its appropriation by women devalued it in male eyes. She argues that this makes Gulliver an object of female consumption and desire, and therefore 'what might be called the ultimate object of female "consumption" – the dildo'.[144] The most famous contemporary work on masturbation, the *Onania*, referred to female masturbators euphemistically as 'pretty Scholars' or 'fair Pupils', drawing precisely on this context of polite scientific study.[145]

In 1745, *The Longitude Discover'd* was republished within a set of poems entitled *The Muse in Good Humour*, appearing directly after another poem entitled *Chloe's Play-Thing: A Tale*, which took the active female sexuality we have been discussing a step further.[146] The poem describes how

> Chloe, a Nymph divinely fair,
> With portly Mien, majestick Air,
> Had often us'd ('twas then the Fashion)
> An Instrument for Titillation;
> So oft, no Lady in the Land
> Was more expert at Slight of Hand ...
> Cloy'd with the Play-Thing, not the Play,
> *Chloe* resolves another Way;
> Ten thousand Methods does explore,
> Experiments not known before.
> Invention racks, in Hopes to find
> A Thing more pleasing to her Mind.
> No *Philomath* e'er pump'd so hard,
> To gain the Longitude-Reward.[147]

Thus, the search for the longitude became part of extreme female sexual empowerment, removing all necessity of a male presence, either intellectual or physical. Indeed, at the end of the poem when Chloe eventually decides that a real man might be a possible alternative, she is disappointed by this 'bauble'[148] in comparison to her 'play-thing'. As readers of the poem, most likely male readers, we are made voyeurs in this impolite spectacle.[149]

For contemporaries, there were no greater voyeurs or proponents of the worst kinds of sexual licence than the infamous Medmenham monks, led

144 Armintor, 'Sexual Politics', p.621.
145 Wagner, 'Veil of Medicine and Morality', p.180.
146 The two poems are followed by another entitled '*Arbor Vitæ*, or the Tree of Life' (*The Muse in Good Humour*, pp.135–9) which makes the obvious botanical metaphor of the penis as tree, the kind of scientific metaphor discussed by Wagner and Peakman.
147 'Chloe's Play-Thing', pp.126–7.
148 'Chloe's Play-Thing', p.130.
149 Wagner, 'Veil of Medicine and Morality', p.180.

by Francis Dashwood (1708–81).[150] This group features in almost every historical account of eighteenth-century sexuality, with special attention paid to the obscene *Essay on Woman*, a parody of Alexander Pope's *Essay on Man*, probably by Thomas Potter (1718–59) but attributed to the radical journalist and politician John Wilkes (1725–97). This poem drew precisely on the questions of hoop petticoats, female sexuality, and their relationship to longitude and latitude, which we have discussed so far. It tells how

> He who the hoop's immensity can pierce,
> Dart thro' the whalebone fold's vast universe,
> Observe how circle into circle runs,
> What courts the eye, and what all vision shuns,
> All the wild modes of dress our females wear,
> May guess what makes them thus transform'd appear.
> But of their cunts the bearings and ties,
> The nice connexions, strong dependencies,
> The latitude and longitude of each
> Hast thou gone through, or can thy Pego reach?
> Was that great Ocean, that unbounded Sea,
> Where pricks like whales may spout, fathom'd by Thee?[151]

What is interesting is how contemporary popular disgust seems to have been as much at the parody of Pope's elegant poem as at the explicit nature of the work. Female sexuality was particularly transgressive in this literary setting.

'My being a Woman, excludes me'

There was, indeed, at least one woman who was confident enough of herself and her ideas to propose a solution to the longitude problem, and to pursue a judgement on it from the Commissioners of Longitude. Jane Squire was publicly vociferous about her ideas in the 1730s and '40s, writing repeatedly to different Commissioners, asking for a Board to be called, and commenting specifically on her position in this process as a woman. The noted bluestockings Elizabeth Carter (1717–1806) and Catherine Talbot (1721–70) corresponded about Squire's proposal the following year.[152] Her

150 Hogarth, *Sir Francis Dashwood at his Devotions* (late 1750s); oil on canvas, 1,200 x 876mm, Private Collection; see Gowing, *Hogarth*, p.50.

151 Quoted in Wagner, *Eros Revived*, pp.54–5. 'Pego' was also the contemporary term for Burma, making this a particularly colonial joke.

152 Carter, *Letters*, i, Carter to Talbot 13 July 1743, i, p.35. My thanks to Michael Ashby for showing me this letter.

Figure 72: Binding from Jane Squire, *A proposal for discovering our Longitude* (London, 1742); leather and gold embossing, 130 x 204 x 23mm, Royal Museums Greenwich, London.

pamphlet was, surprisingly, one of the most owned pamphlets on longitude, appearing in multiple published catalogues of a library or book collection from the 1740s onwards, and quite often the only longitude pamphlet therein.[153] This makes her potentially better known to contemporaries than any other longitude proposer. Some copies of the publication feature particularly beautiful binding that references the imagery used in her proposal (Figure 72). This must have been commissioned by Squire herself, given the niche nature of her work, and therefore shows her attempt to make her book a polite object.[154] After her death in 1743, she was tellingly described in an

153 Examples of such catalogues include *A catalogue of all the English books*; Baker, *A catalogue*; Cater, *A catalogue*; and White, *A catalogue*.

154 Including copies at the William L. Clements Library, Michigan and the Caird Library, Royal Museums Greenwich.

obituary in the *Daily Post* as 'a Lady excellently well vers'd in Astronomy, Philosophy, and most Parts of polite Literature'.[155]

She published *A Proposal for Discovering our Longitude* in both English and French in 1742, including a long series of correspondence with various public figures dating back to 1731.[156] In letters to Sir Thomas Hanmer (1677–1746), the only Commissioner named in the 1714 Act who was then still alive, Squire questioned the motives of the Commissioners, saying, 'it seems to me very easy to demonstrate the Longitude, if it is desired; but impossible to determine, whether it be really desired or no'.[157] We might think about 'desire' and the complex relationship of this to a female proposer that the discussions above have highlighted.[158] Indeed, in her second letter to Hanmer, she answered at length criticisms that she had heard made against her presumption in making such a proposal, saying,

> The first is, that Mathematicks are not the proper Study of Women … but to count, to measure, &c … are so naturally, the Properties of every reasonable Creature; that it is impossible to renounce them, and preserve that Honour … this overvalued Science, seems to me the lowest Business I have proper to me … to study the Law of God Day and Night, is my proper Business; Philosophy my Amusement, and Mathematicks my Play-things … I do not remember any Play-thing, that does not appear to me a mathematical Instrument; nor any mathematical Instrument, that does not appear to me a Play-thing: I see not, therefore, why I should confine myself to Needles, Cards, and Dice.[159]

She presented her mathematical research into the longitude as easy, and as equally appropriate an object of study for her as any man, indeed more appropriate than the 'Needles, Cards, and Dice' that she rejected and some of which we have discussed above as part of the polite sphere. Again, we might note her use of the term 'play-thing' in relationship to Chloe's use of that term above, and what this might have caused her audience to read into her assertion of intellectual independence.

In his reply, Hanmer interestingly invoked the kind of public discussion of science that we have discussed, saying, 'Your Appeal … lies to the World, you must publish your Thoughts and if they carry that Demonstration which you think they do, Mankind will greedily receive them'.[160] He implied,

---

155 *Daily Post* (13 April 1743), Issue 7,363.

156 Journals in a similar style were a popular form of polite leisure activity. See Brewer, *Pleasures of the Imagination*, pp.107–10.

157 Squire, *A proposal*, p.18.

158 The word 'desire' also refers explicitly to the wording of the 1714 Act.

159 Squire, *A proposal*, p.18.

160 Squire, *A proposal*, p.22.

perhaps, that the public would be more open to a proposal from a woman than would the Board. Certainly, in a later reply, after Squire wrote to him again on hearing of the 1740 Act, Hanmer made a revealing apology for his fellow Commissioners:

> your good Sense I am sure will tell you that you are to expect to lye under some Prejudice upon account of your Sex. Man, arrogant Man, assumes to himself the Prerogative of Science, and when a Woman offers to teach them in any of the abstruse Parts of it they are apt to turn a disdainful Ear.[161]

On the back of this, Squire later wrote directly to the Commissioners, after hearing that the Board had met in January 1741.[162] She associated them with the world of public scientific display, opining that,

> What I have been told was your Resolution ... gave me more Surprize than Mortification, when I heard it first; being, that ... tho' I should demonstrate the Discovery of the Longitude ... I should be no better for it, except I would allow the late Doctor *Halley*, Doctor *Desaguliers*, Mr. *Whiston*, or some other Man, noted in Sciences; to propose my Scheme, as his own.[163]

Not only did she take Hanmer's comment about male priority to heart, but she applied it to figures like Whiston, who were so involved in the public performance of lecturing and experimental display, which she read about in the newspapers.[164] They were, in reality, barely more involved with the Commissioners than was Squire, but her assumption that they were tells us about the ways in which she must have learnt about the public face of natural philosophy.

It is striking how reminiscent of a novel Squire's proposal is, with its narrative built out of epistolary engagements. Contemporary commentators both equated novel reading with female masturbation and saw any cultural production by women as analogous to prostitution.[165] In stepping out of the domestic sphere, women transgressed both physical and intellectual boundaries, putting their minds and bodies in the public domain.[166] We have seen a wide range of discussions, both serious and satirical, in which women

161 Squire, *A proposal*, p.54.

162 This is the meeting on 16 January, at which both Harrison and Whiston were given financial support.

163 Squire, *A proposal*, pp.58–60.

164 Both Whiston and Desaguliers had, for example, performed at Steele's Censorium in the 1710s. See Loftis, 'Richard Steele's Censorium', pp.43–66.

165 Brewer, '"The most polite age and the most vicious"' in Bermingham and Brewer, *Consumption of Culture*, pp.353–8, and *Pleasures of the Imagination*, pp.xxii–iii.

166 Rose, *Feminism and Geography*, pp.13–18.

were presented using the longitude problem as a means to transgress. Squire might not have used her body or her sexuality to assert her confidence in her own abilities, as the exaggerated female characters in the poems do, but she was both brash and forward in intruding her ideas into the male sphere of longitude science. Her ideas were, also, certainly promiscuous if she was not, given the breadth of ownership of her pamphlet. For many, therefore, Squire embodied the impolite problem of longitude.

In 'pumping' for her 'Longitude-Reward', Chloe evoked the element of longing and desire that was bound up with the longitude problem: the longing to achieve the 'great reward' that was so effectively satirised in these sorts of erotica. In her confident pursuit of a reward, Squire both evinced such longing—a dangerous force in women—and embodied it as a promiscuous author. The last word on this relationship returns us to Robert Heath and the *British Palladium*. In 1773 he published 'The Longitude, A Song', which we saw in the last chapter discussed Harrison and Maskelyne as projectors on an equal footing. The final two stanzas present the outcome of their projects as the ruin of longitude herself, represented as a prostituted woman at the mercy of the Board of Longitude, and especially Nevil Maskelyne:

> To a dignify'd Board, for Protection, she fled,
> And pass'd for a Maid – with a pure Maidenhead;
> A little-Man took her his Fortune to make,
> Who is fond of her Charms for dear Profit's Sake! ...
> And now her Embraces quite common are grown,
> She's a Cast-off, and Ch[e]at, and a Prost[i]tute known.[167]

It was the joy in pursuing longitude that Heath satirised in this poem. The link that could be made between the elusive longitude and feminine charms meant that one could become tarnished in the same manner as the other.

167 Heath, *British Palladium* (1773), pp.2–3.

6

A Cultural Instrument

IN AUGUST 1760, an advertisement appeared in the *Public Ledger* from

> EDWARD SHUTER, Magical Optician … [who] will open a large and commodious *Warehouse* … during the short Time of Bartholomew-Fair, where will be seen the most uncommon Variety of the greatest Curiosities that were ever exhibited to public View … A new reflecting Telescope, improved upon the Newtonian Plan, for the Discovery of lost Maidenheads, and the Longitude; which will put an End to all the Perplexities of the Profoundest Mathematicians.[1]

Longitude, here, was made an exhibit in the transgressive social space of Bartholomew Fair, conjoined with the socially crucial, but never politely discussed, question of 'lost Maidenheads', and made visible with the help of a scientific instrument. Very differing levels of constancy were required by eighteenth-century opinion of men and women within marriage, where the wife's sexual fidelity was paramount.[2] This rested on preoccupations, for the husband, of property inheritance and of not seeming a fool, on essentially concerns of object and of appearance.[3] So, if a lost maidenhead could be found, along with the longitude, perhaps Astrophel's problems in the previous chapter could be solved by re-possession both of his wife and of his natural philosophical instruments: his 'Trump'ry'?

1 *Public Ledger* (23 August 1760), Issue 193.

2 See for example Dabhoiwala, *Origins of Sex* and Valdimir, 'Patterns of Sexual Behaviour' in Boucé (ed.), *Sexuality in Eighteenth-century Britain*, pp.159–75.

3 Schiebinger, *Nature's Body*, p.196.

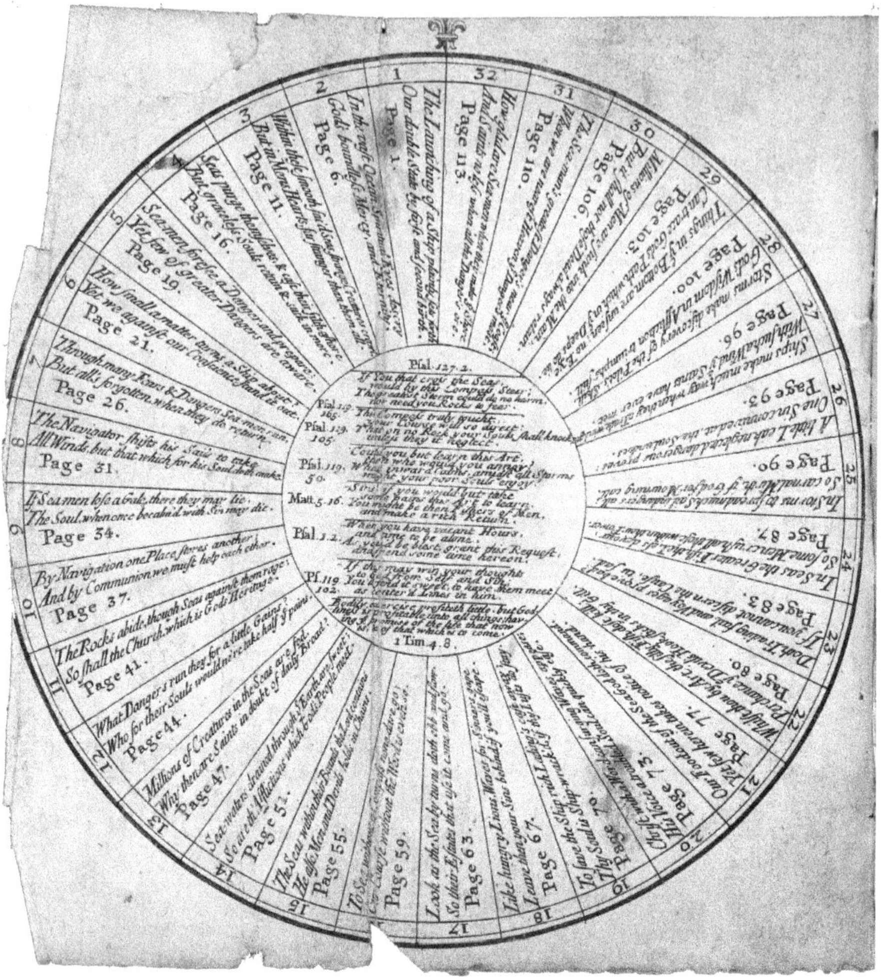

Figure 73: Figure from John Flavel, *Navigation spiritualiz'd* (London, 1721); engraving, 174 x 164mm, reproduced by kind permission of the Syndics of Cambridge University Library [7100.d.1191].

Scientific instruments were a particular site of female transgression. In *The Humours of Oxford*, as Gainlove and Trumore try to negotiate the inconsistencies of female attention, Clarinda, Trumore's beloved, comments to her friend that 'a Beau to a Woman of Fashion, is like a Repeating-Watch to a Citizen's Wife, which she is always striking to let the World know she has such a Bauble belong to her'.[4] Such women take control of interactions between

4 Miller, *Humours of Oxford*, p.39.

the sexes through control, and display, of a complex instrument. One way in which that could be combated was to treat women themselves as instruments. Symbolic relationships were often drawn in the eighteenth century between social interaction, female sexuality, and such instruments as barometers, thermometers, and watches.[5] Indeed, eighteenth-century British culture as a whole can be envisaged as the male object of desire,[6] such that all women in the public sphere become 'commodities'.[7] Thus the two women who observe Hogarth's longitude lunatic become markers of the cultural consumption that surrounded the longitude problem, analogous to instruments.

A range of authors and artists used scientific instruments symbolically as a means of providing the moral boundaries that should contain both men and women in society. They conceptualised this as social navigation, where longitude clearly becomes a boundary. As early as 1664, John Flavel (c.1627–1791) published *Navigation Spiritualiz'd: or, A New Compass for Seamen*, in which he built his exposition of Christian morality around a sailor's 'moral compass', where each point was a line of poetry linking the sea and the soul. It featured such advice as, 'To save the Ship rich Lading's cast away; / Thy soul is Ship-wracked if thy lusts do stay'. This work was republished six times up to the 1720s, and at the beginning of the eighteenth century acquired an image of a compass by which to navigate the text as well as moral dilemmas (Figure 73). Flavel said in his preface to sailors that 'I have endeavoured to cloath Spiritual Matters in your own Dialect and Phrases'.[8] He drew direct parallels between the dangers of navigating a ship at sea and maintaining proper bodily conduct in a tempting world, explaining that

> It is a gallant thing to be able to carry a Ship richly laden round the World: but it is much more gallant to carry a Soul (that rich loading, a Pearl of more worth than all the Merchandise of the world) in a body (that is liable to leaks and bruises as any Ship is) through the Sea of this World (which is as unstable as water, and hath the same brinish taste and salt gust which the waters of the Sea have).[9]

5 See for example Castle, 'Eros and Liberty', pp.156–76 and *The Female Thermometer*. Jan Golinski has built on this to use barometers as markers of complex enlightenment attitudes to science in 'Barometers of Change' in Clark et al., *The Sciences in Enlightened Europe*, pp.69–73.

6 See Brewer, '"The most polite age and the most vicious"' in Bermingham and Brewer, *Consumption of Culture*, p.349. In the same collection, Richard Sorrenson has carried out the only consideration of Harrison within the context of cultural consumption, but emphasises only the state as the market. There are also factual errors in his article, 'The state's demand', pp.263–71.

7 *London Spy*, p.67.

8 Flavel, *Navigation spiritualiz'd*, A4v.

9 Flavel, *Navigation spiritualiz'd*, A5v.

He compared the four points of the compass to four main moral tenets—'making God our North; Christ our East; Holiness our South; and Death our West Points'[10]—and described 'our Memory, that is the Box in which this Compass must be kept'.[11]

In two prints from 1765, Robert Dighton (1752–1814) made similar play with a different type of compass, the kind for drawing lines or boundaries on a map, which we remember one of Halley's navigational graces also holds in his 1702 map. Dighton's *Keep Within Compass* engravings (Figures 74 and 75) are, essentially, versions of *A Harlot's* and *A Rake's Progress*, in which the man and woman are physically contained within the legs of the compass as a means of controlling their morality.[12] The neat, demure woman within her compass (Figure 74) holds a book entitled 'The Pleasures of Imagination Realised', referencing a poem published by Mark Akenside in 1744 on the need for intellectual rather than sensual beauty. She is abjuring the gambling, whoring, drinking, and (inevitable) punishment shown in the surrounding scenes and is following the advice in the accompanying poem to

> Instead of Cards my Fair-one look
> (I beg you'll take it kind)
> Into some learned Author's Book,
> And cultivate your mind …
> When Women once o'erstep the bounds
> Of decency and cares,-
> A crowd of folly quick surrounds,
> And nought but woes she shares.

We might note, however, that she also shares her 'moral compass' with a gambling dog, a standard representation of female sexuality; it is contained here, but nonetheless active and present. Her male counterpart (Figure 75) draws direct relationships with Flavel's sailors. Within his compass, he avoids gambling, whoring, shipwreck, and prison, following eminently naval advice to

> Attempt according to your strength
> Close by the Shore to keep
> 'Tis safer than to hoist up Sails
> And plunge into the deep.[13]

10 Flavel, *Navigation spiritualiz'd*, A6r.

11 Flavel, *Navigation spiritualiz'd*, A8r.

12 Compasses also reference Masonic ideas. See Soulier-Détis, *'Guess at the Rest'* on Hogarth's use of such imagery.

13 A glance through Dighton's other works shows that such progresses remained a theme. The women in his *Twelve Months* series (c.1781) are all keeping well within compass, whereas the man in *The Prodigal Son* set (1792) is doing precisely the opposite.

Figure 74: Robert Dighton, *Keep Within Compass, Prudence Produceth Esteem* (1765); mezzotint, 400 x 280mm, courtesy of the Lewis Walpole Library, Yale University.

Strikingly, these images were then re-used as decoration for domestic crockery from the 1770s, as were a number of Dighton's other works.[14] We can see these ideas of instruments as a means of creating social and moral boundaries moving out, physically, into the social space, mixing with the polite instruments that they referenced. In 1740, *The German Spy* discussed at some length an imaginary thermometer that could measure the emotions of people in different situations and particularly when reading different sorts

14 Such as the plate featuring *Keep Within Compass, Industry Produceth Wealth*, produced by John King Knight, about 1850; Staffordshire transferware, 135mm diameter, Private Collection.

Figure 75: Robert Dighton, *Keep Within Compass, Industry Produceth Wealth* (1765); mezzotint, 410 x 280mm, courtesy of the Lewis Walpole Library, Yale University.

of literature. The author ended by commenting on the myriad uses of such an extraordinary instrument, pointing out how

> it might be for the good of the Public to recommend it earnestly to that illustrious Body of Enquirers into the Wonders of Nature and Art, the R[oya]l S[ociet]y ... to spend some of the Hours they frequently employ upon Matters of mere Speculation ... and it might as well deserve a *Præmium* being set upon it as the Longitude.[15]

One particularly apt broadsheet from 1729 made ample use both of instruments as moral directors and of the symbolic social boundaries that

15 Gentleman, *German Spy*, p.146.

could be created with lines of longitude and latitude. *A New and Exact Map of Toryland, with the Dangerous Rocks and Shoals of all the Jacobite Islands Lying in the same Parallel n^{th} ye Red Sea whose Latitude is 1688, and Longitude 1714* put in visual form many of the ideas of political and religious projecting discussed in Chapter 3 and mapped these with a compass and imaginary grid (Figure 76). To the centre bottom of the map, 'the Jacobite compass rectified' points at all the dangerous directions to which Jacobite sympathies might lead. It orientates the whole map by supposed key figures and features of Tory government including [Bishop Francis] Atterbury (1663–1732), [Jacobite politician James Butler, Second Duke of] Ormond (1665–1745), and [infamous Anglican preacher Henry] Sacheverell (1674–1724), alongside Popery, Superstition, and 'The Devill & all his works', with the Pretender as the north point of the compass. This reminds us forcibly of Flavel's 'moral compass' for sailors. At the centre of the map itself lies 'Toryland', which features counties such as 'Absolution', 'Arbitrary Government', and 'Superstition'. The ships that sail around its coast might easily be wrecked on the 'Passive Obedience Rocks' or the 'Coast of Lost Liberty'. Surrounding 'Toryland' to the west and south is the 'Pretender's Channel' and the 'Pretender's Islands', which include 'Fines and Imprisonments', 'No Tolleration', and 'Loss of Public Credit'. To the east of 'Toryland' lies 'Part of Whig Land', set up as its obvious opposite, where counties include 'Toleration' and 'Parliamentary Right', and the coastline features 'Protestant Point' and 'Hannover Succession Rock'. The map is edged with longitude and latitude scales and the title describes it as showing a location 'whose Latitude is 1688, and Longitude 1714'. Thus, the political map is bounded by a grid created by the Glorious Revolution and the Hanoverian Succession. Yet the Longitude of 1714 might also refer to the Schism Bill, and to the Longitude Act itself.

In 1765, Jonas Hanway (1712–86) in his *Thoughts on the Importance of the Sabbath* used this idea, further, to compare a man to watch-work, saying,

> in whatever character we examine the affairs of human life, *man* may be considered ... as a *time-keeper*, not free of imperfections. So far as intellectual and mechanical powers can be compared, it is by the use of his time, the longitude and latitude of all his thoughts, words, and actions are to be determined.[16]

He not only equated man with an instrument, a link that so many of our authors above made with women, but he then drew the boundaries of his intellectual abilities with lines of longitude and latitude. In *Gulliver's*

16 Hanway, *Thoughts*, p.51.

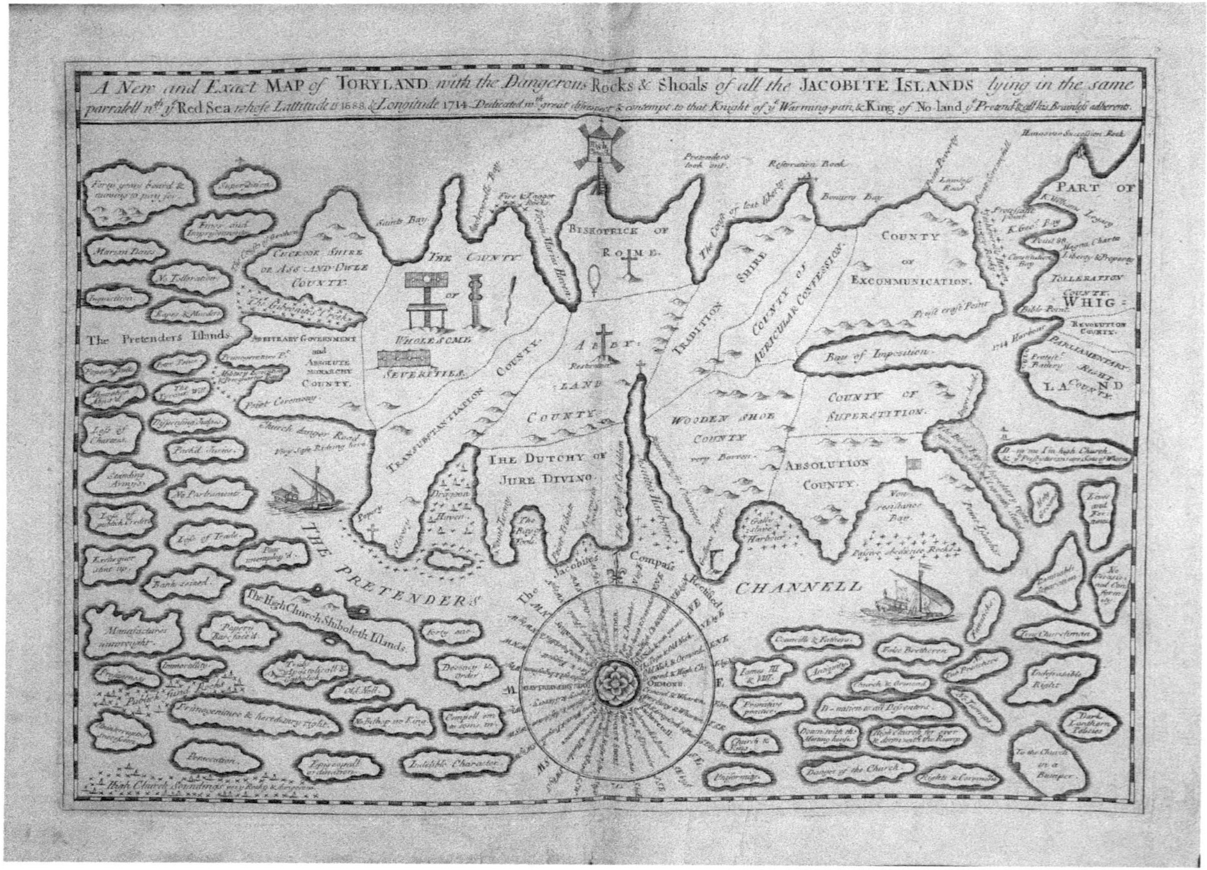

Figure 76: Anon., *A New and Exact Map of Toryland, with the dangerous Rocks and Shoals of all the Jacobite Islands lying in the same Parallel nth ye Red Sea whose Latitude is 1688, and Longitude 1714* (1729); engraving, 565 x 576mm, Bodleian Library, University of Oxford: fol. DELTA 755 (1).

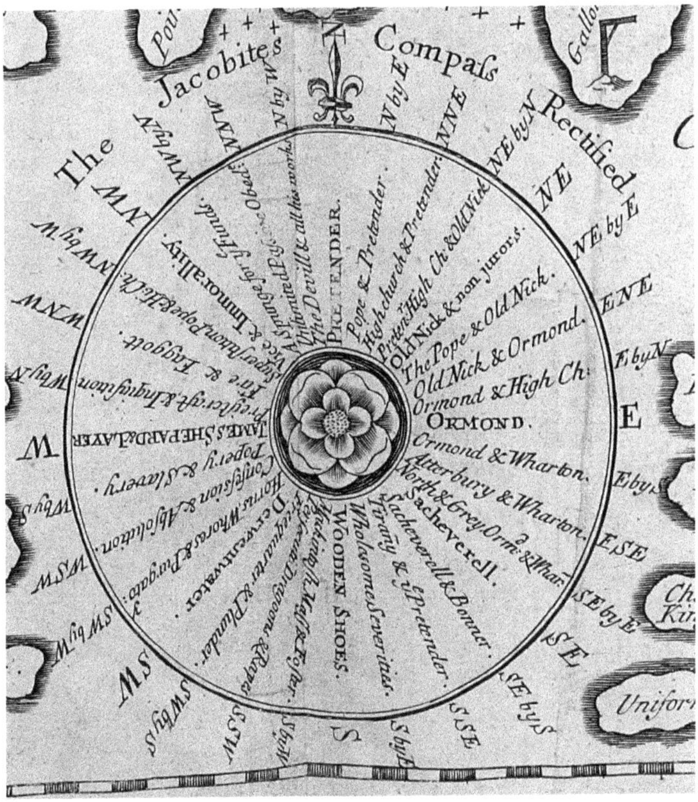

Figure 76: detail.

Travels, Swift had similarly made Gulliver's changing size, and therefore his status and vulnerability, visible through his ability to hide his watch, or be mistaken himself for a piece of clockwork.[17] This is the same setting in which Armintor associated him with a pocket microscope and, therefore, a dildo: when powerless Gulliver became an instrument, just like a woman. This puts an added spin to Astrophel's parting comment on finding his wife amongst his instruments, flushed and aroused on the floor. In commenting that, 'Whether a Longitude you found, / Fit for a Judgement so profound, / I know not; -- but to end my Care, / Here's Latitude – I see – to spare',[18] he drew his boundary of acceptable behaviour, the latitude line that she had most definitely crossed, while she drew the longitude line against which she wished to measure him.

17 Swift, *Gulliver's Travels*, pp.28, 93. Schaffer links questions about valuing clockwork to debates about Orffyreus' perpetual motion wheel in 'The Show That Never Ends', p.182.

18 Author of The Deluge, *Longitude discover'd*, p.12.

'Now or never, Mrs. Longitude'

From 1759 to 1767, Laurence Sterne published nine volumes comprising one of the most famous novels of the eighteenth century, *The Life and Opinions of Tristram Shandy*, a work similarly obsessed with drawing lines.[19] The story revolves around two brothers, Walter and Toby Shandy, and their differing sexualities and temperaments. In the language he chose to discuss these, Sterne picked up on precisely these ideas of longitude and instruments. Each marks important points in the lives of one of the brothers, and both are a perfect example of how popular erotica trickled up into high literature.[20] Tristram's father Walter's story stems from discussion of the family clock. The novel opens with Tristram's mother asking, 'Pray, my dear … have you not forgot to wind up the clock?' at the very moment of his conception.[21] This outburst is explained by his father's regular habits in life:

> he had made it a rule for many years of his life, —on the first Sunday night of every month throughout the whole year … to wind up a large house-clock, which we had standing upon the back stairs … he had likewise brought some other little family concernments to the same period … it so fell out at length, that my poor mother could never hear the said clock wound up, —but the thoughts of some other things unavoidably popped into her head—& vice versa.[22]

Thus, the Shandys' sex life is tied to the running of the clock. By contrast, a major theme of the novel is Uncle Toby's innocence and chastity, and his failed courtship of his neighbour, Widow Wadman. References to longitude appear at two points in the novel, both of which deal with questions of Toby's temperament. The first discusses his tender heart in the context of an infinitesimal social line that should not be crossed, for 'before an affliction is digested, —consolation ever comes too soon; —and after it is digested, —it comes too late … there is but a mark between these two, as fine almost as a hair … my uncle Toby was always either on this side or on that of it, and would often say, He believed in his heart, he could as soon hit the longitude'.[23] The second is in the context of his wish for children, when 'I think the procreation of children as beneficial to the world, said Yorick, as the finding

19 Hogarth provided frontispieces for the second edition of Volumes 1 and 2, and the first edition of Volumes 3 and 4.

20 Wagner, *Eros Revived*, pp.303–9.

21 Sterne, *Tristram Shandy*, p.35.

22 Sterne, *Tristram Shandy*, p.39.

23 Sterne, *Tristram Shandy*, p.223.

out the longitude'.[24] We might see the longitude as helping to mark out Toby's implied impotence, but also his keen awareness of appropriate social behaviour, as opposed to his brother's sexuality, which runs like clockwork.

What is most remarkable is the way in which these two elements as markers of the brothers' sexualities were picked up and developed in a number of satires on Sterne's work. In 1760, *The Clockmakers Outcry* was published, making a marvellous satirical complaint that the tale of his father's clock-winding had resulted in 'bringing the works of our fraternity into disgrace' by turning clocks into sexual objects.[25] The authors opined that

> no modest Lady now dares to mention a word about winding up a clock … The common expression of street-walkers is 'Sir, will you have your clock wound up?' … Reputable … clocks that have flourished for ages, are ordered to be taken down by virtuous matrons, and be disposed of as obscene lumber, exciting to acts of carnality.[26]

Men and women, it goes on, can now express sexual interest by winding their watch while looking at a preferred member of the opposite sex, and even 'sober and well-regulated clocks are treated as the alarms of lust, as veteran bawds'.[27] Likewise in 1766, a spoof Volume 9 was published before Sterne's own appeared, in which, as part of their attack on Widow Wadman, Uncle Toby and his assistant Trim consider the state of Toby's body in the language of fortification. They decide that his 'horn work' is out of repair, so need to call in a doctor, and this doctor is then used by the widow to try to discover whether Toby is impotent. Tristram comments that 'Mrs *Wadman* could not for her soul but receive Dr. *Querpo* in a very languishing manner, as he put her so strongly in mind of my uncle. This was perfectly making a house-clock of him'.[28] Thus, the clock is made a transgressive object through Sterne's associations.

Most brilliantly, in 1761 *The Life and Opinions of Bertram Montfichet* was published as an obvious satire on the first volumes of *Tristram Shandy*. In this work, Toby is exaggerated in the character of Uncle Dick, who has been raised as a girl until seeing the family maid naked and realising his bodily differences.[29] Bertram explains that Uncle Dick was so innocent in mind that 'he carried his maidenhead with him to the grave' probably because spending

24 Sterne, *Tristram Shandy*, p.561.
25 *Clockmakers Outcry*, p.14.
26 *Clockmakers Outcry*, p.42.
27 *Clockmakers Outcry*, p.43.
28 Sterne, *Tristram Shandy*, p.27.
29 This fits with what Friedl has written about the use of male clothes by women to assert economic status and reduce sexual vulnerability. See 'Passing Women' in Rousseau and Porter, *Sexual Underworlds*, p.234.

so long in petticoats made him averse to women.[30] After announcing his gender change to his family and clothing himself in a sailor suit,

> my uncle *Dick* arose, full of thought, how he should occupy himself in a manner worthy of his new acquired manhood … We are indebted to him … for several very ingenious processes, in order to facilitate the discovery of the Longitude and perpetual Motion … one night in bed, whether asleep or awake, he was not certain; but the longitude appeared to him … as tractable as any wife during the honey-moon … [but] to recall and fix the idea of his dear longitude, when the light of the morn opened his eyes, he struggled in vain.[31]

Dick tries lying at every different angle to remember his discovery, but none suit his body, and he ends exclaiming, 'Well, now or never, Mrs. Longitude; pray, did not I ferret you last night out of your lurking hole, and did not I afterwards, besides taking your exact dimensions, survey minutely all your attitudes?'[32] Longitude becomes the means of Dick seeking and missing his 'new acquired manhood' and as such is personified as a woman 'as tractable as any wife during the honey-moon' in his dreams but struggled for in vain when awake. Dick ends, further, by singing his own loose variation on the Scriblerian 'Ode to the Longitude' discussed in Chapters 3 and 4, opining,

> The Longitude! Who hit it?
> I honest Dick Montfichet,
> In a dream!
> In a dream!
> The Longitude! Who miss'd it?
> I poor Dicky Montfichet,
> Quite awake!
> Quite awake![33]

Tristram Shandy was celebrated by contemporaries as a penetrating discussion of social mores. In one complimentary work, published the year after this satire, John Gordon (1726–93), Archdeacon of Lincoln, commented that

> I heartily wish the ingenious *Shandy*, in that new map of the human mind, which he is publishing, had given a fuller and more accurate chart … I care not if he had even a part of the reward assigned him, which is voted by Parliament for the discovery of the Longitude: I mean however only so much of it, as will not be allotted to Mr. *Harrison*; who perhaps deserves it all.[34]

30 Montfichet, *Life and Opinions*, i, pp.122–3.
31 Montfichet, *Life and Opinions*, i, pp.126–8.
32 Montfichet, *Life and Opinions*, i, p.130.
33 Montfichet, *Life and Opinions*, i, pp.153–4.
34 Gordon, *Occasional thoughts*, p.142.

Thus, Sterne's own success in mapping the contemporary mind was compared with the difficulty of finding longitude, his novel was made equal with John Harrison's work on timekeepers, and the language of longitude and of instruments which he drew so effectively from popular erotica was employed to compliment his own success.

'Mr. Harrison's size is little enough for the longitude'[35]

As Sterne penned the complex amours of Uncle Toby, John Harrison was deep in arguments with the Board of Longitude over how to win the reward. Where do these discussions of polite and impolite science, instruments as cultural and moral barometers, and longitude as a social boundary leave him? Harrison's problems were somewhat different from those of Jane Squire in his interactions with the Board.[36] He was a prized holder of instrumental knowledge, and he had gained personal access to the Board. Indeed, one of Squire's letters in 1742 seems to be responding to a garbled report of their meeting.[37] He had *the* instrument in the middle of the eighteenth century, and priceless instrumental knowledge. This placed him physically within the polite world of contemporary science and required him to interact with the Commissioners there. Harrison was well aware that he had to play a part in this setting, commenting in one publication that 'one would think, a Minstrel at the Play-House is much better off, save only that this has been my great Delight'.[38] Like the transgressive women in longitude erotica he was different because of this delight, and because he displayed his difference. In some ways he was like a woman, less visible than his instruments, and yet measured by them. The difficulty was in knowing how he was being measured and what he should and should not display. This is what the Harrisons continually discovered in their discussions with the Commissioners; again and again, they complained that 'there is two words'[39] in all the Board's requirements which they did not understand and, therefore, could not comply with. Those words were 'experimental exhibitions', which were laden with the social actions necessary to perform those both in person and in print. They lie behind their discussions over the means of inscribing

35 Hatton, *Introduction*, p.370.

36 Brewer has discussed practising self-reflexion in a diary as central to polite behaviour in *Pleasures of the Imagination*, pp.107–10. We might think about both Harrison and Squire as performing this practice.

37 Squire, *A proposal*, pp.54–8.

38 Harrison, *Description*, p.52.

39 *HJ*, p.90.

Harrison's designs on paper, and behind the language of projecting and madness that surrounded the longitude problem.[40]

The phrase 'experimental exhibitions' first entered the fray after the 1765 Act, when the Commissioners and the Harrisons began to consider his disclosure. In addition to the drawings and explanations discussed in Chapter 1, the Commissioners required that Harrison should

> give a further explanation by word of mouth, and experimental exhibitions where judged necessary, of the said Principles, and of everything relative to the construction of the said Watch; producing the same; taking it to pieces; & answering, upon Oath, to every question proposed by the Board.[41]

In response to this, the Harrisons 'endeavoured to shew the absurdity ... as there was no knowing what might be demanded from it',[42] for 'he might never know when he had done from the words, Experimental Exhibitions'.[43] Indeed, it was clearly the particular word 'exhibitions' to which Harrison took exception, as, in a short interlude when they seemed to agree on terms, he had no problem with the re-worded requirement from the Commissioners that he should

> take the Timekeeper to pieces before them and explain all its parts. That he shall go regularly through the Explanation of the whole Instrument and if in this process there shall happen to be anything which those Gentlemen do not understand then Mr Harrison shall be put upon his Oath and explain the same.[44]

A letter Harrison submitted to the Board showed clearly that his fear of the term 'exhibition' was that it required him 'to instruct People that I know nothing of, and such as may know nothing of Mechanicks; and if I do not make them understand to their satisfaction, I may then have nothing!'[45] Likewise, in his published *Case* of 1767 he explained how 'he was required by *Experimental Exhibitions* to explain whatever the Commissioners and whom they should appoint, might desire, and they refused to limit themselves as to those Exhibitions, either in Point of Time, Nature, or Number'.[46] Demonstrations of clockwork played an important role among the contemporary spectrum

40 This also seems to be a new usage by the Commissioners. The words do not appear together in any other texts on ECCO.

41 *HJ*, p.80.

42 *HJ*, p.81.

43 *HJ*, p.83.

44 *HJ*, pp.85–6.

45 *HJ*, p.84.

46 Harrison, *Case* (c.1767), p.2.

of exhibitions, with clockwork objects moving between polite and impolite spaces, regularly appearing at fairs, but also at pleasure gardens.[47] Harrison clearly saw the performative aspect of experimental displays that 'exhibitions' would require him to carry out in order to make his watch visible to the uninformed, and which he felt himself incapable of performing.

Part of the problem was that there was by now intense public interest in the Harrisons' timepieces, as we saw in Chapter 4. The relationship between the Harrisons and the Board was coming more and more under scrutiny, and therefore both had to engage with the polite, public discussion of the longitude debate that Squire so clearly understood in her references to Whiston. The *Universal Magazine* itself discussed the passage of the Act in 1763 to encourage Harrison to disclose his principles, and in 1765 it gave an abstract of his *Narrative* and the terms of the new Act.[48] In his *General Magazine* of 1764, Benjamin Martin also published a letter concerning Harrison, commenting how

> some imperfect accounts having already appeared in the Newspapers about the Results of the Trials of Mr. *Harrison's* Longitude Time-keeper ... it has been thought proper, in some measure, to satisfy the impatience of the Public, till a board of Longitude shall be held ... to publish the following authentic and plain Narrative of some Experiments.[49]

Martin was clearly aware of the interest of his public in the natural philosophy behind Harrison's watch and thus published these 'private' experiments until official demonstrations could be ordered by the Board. He understood the polite fascination in Harrison's unique instrument. The Commissioners were clearly also aware of this public interest. We saw in Chapter 4 how they agreed to publish the minutes from 1765, in order 'that the World might not be imposed on',[50] with all the connotations of projecting and madness which were there discussed. Such publication not only showed the Board's concern to engage with public interest in the fate of Harrison's solution but, further, the minutes published dealt particularly with this vexed question of 'experimental exhibitions'.

Harrison also had experience of having to perform in his workshop, which he would have to do as part of the disclosure, as his timepieces became objects on the circuit of London sightseeing. His watch was the only specific instrument to be cited by *The London and Westminster Guide* of 1768 in its

---

47 Altick, *Shows of London*, pp.56–71.

48 *Universal Magazine* (Volume 33, 1763); Harrison, *Narrative*.

49 Martin, 'Biographia Philosophica' in the *General Magazine*, xiii, p.336.

50 *HJ*, p.91.

discussion of London mechanics.[51] Harrison appears to have struggled with the social niceties these visits required. Bengt Ferrner visited Harrison during his stay in London, and it is noteworthy that he records no real conversations or social interactions with him, compared to the other makers—Dollond, Bird, and Short—with whom he became socially intimate. He mentions Harrison five times in the course of his 'Journal', mostly giving only a factual statement of seeing the clock, discussing others' opinions on it, or focusing on the other people of interest who were also present at his visit. On one occasion only, he records that Harrison told him he would go to sea the following year and had received money from the Board.[52] Bernoulli, likewise, discussed two occasions on which he saw Harrison's instruments, which he praised and described at some length. One was at the Greenwich Observatory and one at Harrison's house in Red Lion Square, but Bernoulli specifically said that he did not see Harrison.[53]

Shops were important as 'semi-public space in which visitors conversed with makers' in the eighteenth century.[54] They stood between the workshop and the street, and thus provided a polite area between public bustle and commotion and the busy world of mechanical construction. Instrument makers were 'trading in mathematics and natural philosophy—their material products were supposed ultimately to deliver more than materiality to their customers—so they needed to present a "polite" image to the world'.[55] The Harrisons did not have a shop, and did not sell instruments commercially, nor were they based in the same Fleet Street area as most clockmakers. John Harrison seems to have taken on some commissioned work in the 1750s, probably to supplement his income, but in essence he worked exclusively for the Board.[56] He therefore had no space between his private workshop and the public street, no polite setting for any 'experimental exhibitions'. We might think of Harrison letting the Commissioners into his workshop, as well as into his secrets, as analogous to the impolite behaviour of the women we have discussed above. They would step over the polite boundary and into his vulnerable private space. Harrison seems to have realised this himself, and

51 *London and Westminster Guide*, p.xv.

52 Ferrner, *Resa i Europa*, p.330.

53 Bernoulli, *Lettres Astronomiques*, pp.96, 110.

54 Bennett, 'Shopping for Instruments', p.374.

55 Bennett, 'Shopping for Instruments', p.388.

56 See Andrewes, *Quest*, p.220. At the Commissioners' meeting in 1746, Harrison explained that 'the Difficulty in the Contrivance & the Nicety in the execution of many parts of these new designs have so entirely ingrossed his time & thoughts for many years past as to render him quite incapable of following any gainfull employment'. Meeting 4 June 1746, RGO 14/5: 14.

to have attempted to turn it to his advantage when, as part of his wrangles with the Board over who should attend the 'experimental exhibitions', he commented that 'they might appoint who they pleased, But he would let come into his House who he pleas'd'.[57]

Shop visitors were also seeking particular kinds of instruments. They wanted 'the original, ingenious, or ornate piece, an instrument worth a special visit'.[58] Harrison had this by definition: he had *the* instrument worth a special visit in the middle of the eighteenth century. It was described by one pamphleteer as 'the most ingenious, and best Clock that ever was … made, for this purpose',[59] and the antiquarian William Stukeley, whom we have already met praising Harrison's genius in the last chapter, described how 'the sweetness of [its] motion … cannot be sufficiently admired'.[60] Yet we have seen how instruments occupied a complex cultural role as measurers of morality.[61] Harrison faced the tension between his watch as an object of *vertu* and as an instrument to be made practicable and replicable. Many of the figures who supported Harrison on his first move to London and interactions with the Board were collectors interested in the beautiful and unique object: Hans Sloane, Martin Folkes, the Earl of Macclesfield (1695/7–1764). It is significant that the men with power in the early years of the Board tended to be such virtuoso men, rather than the naval members.[62] Yet the Board equally required the watch to be replicable, and other instrument makers questioned that it could be, especially when Harrison repeatedly asked the Board for access to it in order to make another. Kendall and Mudge both commented that 'it is very strange that a Man cannot make something from something, when he has made it from nothing'.[63]

Harrison claimed directly that 'it is not in the power of any Man to make and adjust such Timekeepers till he has either had experience acquired by his own labour, or communicated to him by me'.[64] His assertion that the watch essentially embodied his knowledge, and that only the process of making it could allow other makers access to that knowledge, never seems to have been accepted or even recognised by the Board. His assertions link directly with

57 *HJ*, p.82.

58 Bennett, 'Shopping for Instruments', p.376.

59 Costard, *History of astronomy*, p.249.

60 Stukeley, *Family Memoirs*, ii, p.298.

61 Pomian has likewise discussed the complex place that curiosities such as medals held between dilettante social control and the search for self-improvement. See 'Medals/Shells=Erudition/Philosophy'.

62 On the culture of virtuosity in this period see Hanson, *English Virtuoso*.

63 *HJ*, p.98.

64 *HJ*, p.109.

arguments for the self-consciously embodied nature of artisanal knowledge, and the need to bring the social, physical dimension of this knowledge into play alongside the visual and intellectual.[65] Harrison was his watch, in terms both of the materiality of his knowledge and of the public view of him. We might again, therefore, compare him to the women who used the longitude problem as a means of asserting their bodily power in erotic literature. Words like 'artful' and 'subtle' were regularly used to discuss feminine sexual wiles.[66] Harrison's watch depended on similarly artful skills. It gave him the power to assert his confidence in a polite setting within which he was expected to behave in a certain way, it allowed him to push at the boundaries of that setting, but it also created questions around the status of his knowledge and his ability to communicate it. The mechanical, replicable element of his watch was like the bodily danger of women: its native charms could only go so far.

Thus, Thomas King's (d.c.1796) portrait of Harrison painted in 1765–66 was as much about his clocks as about him (Figure 77). He sits confidently at the centre of the image, in a traditional masculine pose, clasping in his right hand the watch made for him by John Jefferys (1701–54), the background taken up entirely by H3 and his gridiron pendulum. It is, essentially, a double portrait of the man and his machines. By 1767, this image, like our *Rake's Progress* plate, had also been changed for the sadder. In Philippe Tassaert's (1736–1803) engraving (Figure 44), Harrison's face and demeanour have become markedly more refined, polite in fact, but the instruments behind him have shrunk such that he dominates the composition. His hand now makes a soft, elegant gesture towards the engraved image of watch-work that lies on the table, while the Jefferys watch has been replaced by H4 lying on its back on top.[67] The 'Harrison Journal' ends with the painful scene when Nevil Maskelyne arrived at Harrison's house to collect his timepieces and remove them to the Observatory for trials. He and Harrison took some time to agree the wording of a certificate to the effect that Maskelyne had received them 'by all appearance in perfect order and good Condition'.[68] We note the emphasis on appearance, as Maskelyne could not know for sure that the timepieces' mechanisms were in 'perfect order', nor could Harrison give him that guarantee in a manner that Maskelyne could understand, as

65 Smith, *Body of the Artisan*. The subjects of Smith's work are chronologically and geographically separated, but tackle the same questions considered by the Board a century later.

66 Spacks, 'Ev'ry Woman is at Heart a Rake', p.30.

67 Andrew King has discussed changes to Harrison images in 'The Images of John Harrison', pp.538–43.

68 *HJ*, p.113.

Figure 77: Thomas King, *John Harrison* (1765–66); oil on canvas, 1,250 x 1,000mm, Science Museum / Science and Society Picture Library.

the knowledge was in the machines themselves. In the removal process, H3 was damaged, at which point the 'Journal' asserts that 'Mr Maskelyne's hand was … on the Machine'.[69] Maskelyne's hand was on the machine whereas

69 *HJ*, p.114. Harrison made this accusation explicit elsewhere, stating that 'one of [the clocks] … was broke to Pieces *under his careful and ingenious management*', *Remarks*, pp.22–3.

Harrison's was not. In removing these from him, the Commissioners removed his embodied knowledge and the instruments that allowed him to negotiate the public world of science. In the engraving, he therefore no longer holds a watch but points to an engraving, the more acceptable representation of his knowledge. He has become more 'polite', but seems to have shrunk, lacking the confidence in this world that his timepieces, his cultural instruments, had given him to negotiate what he should 'exhibit'.

Conclusion to Part III

HOGARTH HAD BEEN AMONG THE POLITE PUBLIC who visited Harrison's workshop in the early 1750s, and he commented at some length in *The Analysis of Beauty* on differences of appearance and function in Harrison's third timekeeper (H3):

> A clock, by the government's order, has been made, and another now making, by Mr Harrison, for the keeping of true time at sea; which perhaps is one of the most exquisite movements ever made. Happy the ingenious contriver! although the form of the whole, or of every part of this curious machine, should be ever so confused, or displeasingly shaped to the eye; and although even its movements should be disagreeable to look at, provided it answers the end proposed: an ornamental composition was no part of his scheme, otherwise than as a polish might be necessary; if ornaments are required to be added to mend its shape, care must be taken that they are no obstruction to the movement itself.[1]

While Hogarth's concern was with the line of beauty, he recognised that Harrison's concern was not with appearance but with mechanism. Throughout his interactions with the Board of Longitude, and with the wider public sphere, Harrison struggled with the social forms, the appearances, necessary to present his knowledge and his machines in the appropriate way.

Politeness in the eighteenth century, both artistic and scientific, was about appreciating the intellectual over the manual, the aesthetic over the practical, the mind over the body, conversation over intercourse; pushing the latter

1 *Analysis*, pp.61–2.

in each pair to the edges of social interaction. With its status as a soluble and insoluble scientific problem, and the focus of intense public curiosity, longitude provided a space in which to negotiate where the boundaries were between the polite and the impolite, in science and more widely.[2] The words 'experimental exhibitions' summed up how this kind of discussion troubled John Harrison. His confident material knowledge became less real, and less accessible to him, the further he entered the polite world inhabited by the Commissioners of Longitude and the reading public. His 'lesser' bodily, mechanical skills were still as important, just as the women participating in polite scientific discussion were still expected to flirt and entertain their male companions. What Harrison had to negotiate were the boundaries of how these elements could be performed. His unique instrument both allowed him to do this and became a measure of his success.

2 Gilpin, *Essay upon Prints*, pp.227–8.

Conclusion
The Longitude of London

'LONGITUDE' IS NOT THE ONLY WORD that Hogarth's lunatic wrote on the wall of Bedlam. Alongside his map of the world and his various solutions, he also labelled the North Pole and the Antarctic Circle (Figure 1). He signposted the conclusion of our phase in the story of longitude and the new directions to which the Board would turn their attention as Harrison left the story. This book has shown how Hogarth's image was both ordinary and extraordinary: unusual in being a rare image of someone trying to solve the longitude problem yet directing us to a vast network of visual and textual sources that discussed, ridiculed, and analysed it. We have seen how this mass of discussions and imagery shaped the efforts of John Harrison and the Board of Longitude to find a solution 'practicable and useful at Sea' but also at home. As the concerns of the Board and public interest in longitude moved away from negotiations in London from the 1770s, broader dreams of naval victory and voyages of discovery came into play, taking longitude firmly out onto the world's oceans.[1]

In 1773, John Harrison finally came as close as he ever would to winning the great reward. Despairing of their negotiations with the Board, John and William Harrison sought the patronage of the King. With their final watch, H5, in hand they approached Stephen Demainbray (whom we met offering free lecture tickets to ladies in Chapter 5), now astronomer at the King's private observatory at Kew, requesting a trial. After William met George III (1738–1820) at Windsor, a trial took place at Kew from May to July 1772,

1 For longitude and instrumentation at sea see Phillips, 'Making Time Fit'.

which showed the watch to lose on average less than a third of a second per day.[2] Yet the Board maintained their requirement for an official trial at sea. The Harrisons petitioned Parliament but were advised that the Board had acted legally. A second petition therefore simply called for generosity and, with the King's support, Harrison was voted £8,750 by Parliament as 'a further Reward and Encouragement'.[3] He would enjoy his success for only three years, dying in 1776.

For the Board of Longitude, the focus then shifted in two directions.[4] First, to the attempt to create versions of Harrison's watch that could be cheaply made and would remain accurate at sea. Here, Larcum Kendall, John Arnold (1736–99), and Thomas Earnshaw (1749–1829) became the main protagonists.[5] In 1772–75, with Kendall's copy of Harrison's watch (K1) on board, James Cook (1728–79) and the crew of the *Resolution* navigated the Southern Ocean.[6] They finally put paid to the idea of an unknown southern continent reaching to the South Pole, which Hogarth's lunatic's label of 'Antarctic Circle' referenced. Second, the Board put considerable effort into the production of the *Nautical Almanac*, which supported both the lunar distance and timekeeper methods of finding longitude. Sextants entered large-scale production thanks to Jesse Ramsden's (1735–1800) dividing engine for cutting accurate scales, and Nevil Maskelyne, as Astronomer Royal, continued to marshal an impressive number of computers and comparers (both male and female) to complete calculations every year.[7] The Commissioners published a description and figures of Ramsden's dividing engine in 1777,[8] and Maskelyne's concerns over printed accuracy and display shifted to the pages of the *Almanac*.

Proposals continued to flood into the Board suggesting new and updated solutions, but the Board's focus of attention also expanded as the century progressed. From 1774 its brief included navigation more generally, and from 1818 it became involved in the search for the North-West Passage

2 Dunn and Higgitt, *Finding Longitude*, pp.121–5; Baker et al., *The Board of Longitude*.

3 13 George III c.77, Section XXIX, Parliamentary Statutes online.

4 For the wider ongoing story of the Board see Dunn and Higgitt, *Finding Longitude*, pp.126–225.

5 Betts, *Marine Chronometers*, pp.3–72; 'Arnold and Earnshaw' in Andrewes, *Quest*; articles on Kendall, Arnold, and Earnshaw in *ODNB*.

6 Larcum Kendall, '*K1*' (1769); timekeeper, 165 x 124 x 28mm, Royal Museums Greenwich.

7 On Ramsden's development of the dividing engine see Chapman, *Dividing the Circle*, pp.129–33; on the *Nautical Almanac*, computers, and comparers see Croarken, 'Tabulating the Heavens', pp.48–61.

8 Ramsden, *Description of an Engine*.

Figure 78: William Hogarth, *A Rake's Progress, Plate 8* (T. Cook, 1808); engraving, 145 x 175mm, © Trustees of the British Museum.

and the attempt to reach the North Pole. This became the nineteenth century's impractical or impossible scheme, referenced by Hogarth's lunatic on his map of the world.[9] The search for longitude as such, then, never reached a definitive conclusion. The 'great reward' was never gained and the network of scientific, visual, mental, and social concerns that enmeshed longitude discussions simply shifted as the Board's attentions turned to new instruments, papers, and problems. It is apt, then, that by 1808, when Thomas Cook (c.1744–1818) produced new engravings of Hogarth's *A Rake's Progress* (Figure 78), the longitude diagram became almost invisible on the back wall of Bedlam. For longitude had left this metropolitan context.

9 On the later years of the Board of Longitude see Waring, 'Thomas Young'.

Painting the Town

Looking for longitude throughout this book has involved navigating the geography of metropolitan London. We have seen how many of the protagonists in the previous chapters reference specific London spaces in their pamphlets, whether challenging opponents to a trial at the Royal Society, opining at the flood of papers emanating from Grub Street, visiting the madmen at Bedlam, or including Harrison's workshop in the round of fashionable visits that also took in the Tower or Vauxhall Gardens. Longitude experiments took place on Shooter's Hill and Hampstead Heath, in Red Lion Square and the Fleet prison, on a canal in St. James's Park and in the King's observatory at Kew.[10] Meeting at the Admiralty, testing Harrison's watch (H4) at the Royal Observatory, or reporting in Parliament, the Commissioners of Longitude similarly operated within a variety of London sites. Longitude pamphlets were discussed in the coffee houses, in taverns, and at the publishers, while engravings circulated from print shops to libraries and drawing rooms. These are the paper worlds in which our story of longitude largely operated, ephemeral but multiple, known to a wide-ranging eighteenth-century public, referenced, satirised, and understood.[11]

Longitude also took its place in the more permanent painted spaces of London institutions. Two painted references to longitude still exist within the capital and testify to its longer institutional and visual history, enshrining longitude in the ongoing iconography of the city. The year of the Longitude Act, 1714, was also the year in which James Thornhill (1676–1734)—future sergeant painter to the King and father-in-law of Hogarth—finished work on the Painted Hall at Greenwich Hospital: a baroque masterpiece celebrating British maritime power. The ceiling includes a portrait of Astronomer Royal John Flamsteed, who, as we have seen, became a Commissioner of Longitude from 1714. His position of Astronomer Royal had been created by Charles II (1630–85) in 1675, with a new observatory at Greenwich, specifically to support methods for measuring longitude at sea. Flamsteed appears in the south-east corner of the lower hall ceiling with, much like other protagonists in our story, an instrument—his self-designed arc telescope—and an illustration—a dated diagram of an eclipse (Figure 79). By 1714 he had been at work for decades

10 Experiments in St James's Park were discussed by De Saumarez in *Letter to a friend*, p.7. All other examples appear in earlier chapters.

11 For more on the geography of longitude in London see Barrett et al., 'Looking for "the Longitude"'.

Figure 79: James Thornhill, Detail of the ceiling of the Painted Hall, Old Royal Naval College, Greenwich (1714); oil on plaster, © Old Royal Naval College.

on the important work of celestial observation that it was then hoped would help in the search for longitude, and which attracted significant public attention. Thornhill's inclusion of Flamsteed in the ceiling of the Painted Hall celebrated this observing work and particularly his famous prediction of a comet that would appear in 1715.[12] In holding the prominent diagram, Flamsteed again shows the importance of the circulation of ideas on

12 For the history and iconography of the Painted Hall more generally see Johns, 'James Thornhill', particularly pp.180–4 on Flamsteed's role in the iconography. Also Lucas et al., *Painted Hall.*

Figure 80: James Barry, *Commerce, or The Triumph of the Thames* (1777–1801); oil on canvas, 360 x 462mm, © Royal Society for the Encouragement of Arts, Manufacturers & Commerce / Bridgeman Images.

paper right from these early days of the longitude story. He plays his part in Thornhill's painted splendour dedicated to Britain's maritime and commercial success, putting the people, instruments, and papers of the longitude problem firmly at the fore.

By the 1770s, longitude was associated with a very different maritime mural in a very different institution. The potential of accurate timekeepers and the lunar distance method for finding longitude at sea had been established by the voyage of the *Resolution* to the South Pacific. James Cook subsequently appeared as one of four navigators supporting Father Thames in the suite of six murals started in 1777 by James Barry (1741–1806) for the Great Room of the new Society of Arts just off the Strand (Figure 80). Entitled *A Series of Pictures on Human Culture*, Barry intended the paintings to both celebrate and direct the progress of British cultural life, with scenes of modern Britain following three that celebrated Ancient Greek culture. The first scene in Britain shows *Commerce, or The Triumph of the Thames*, in which Cook is accompanied by Francis Drake (c.1540–96), Walter Raleigh (1552–1618), and John Cabot (c.1450–c.1500) as they support Father Thames' majestic progress out to sea, where he is offered produce by the four continents. The Thames and the City of London are tightly conjoined in their commercial success, with the city's coat of arms adorning Father Thames' throne. The future of longitude is here embedded in Barry's celebration of the capital as the centre of British trade and culture, as well as in the figure of the heroic navigator, enabled to expand Britain's view of the world through use of instrument and paper.[13] In 1803, he would add a scroll to the murals referencing the Society's 1792 offer of a gold medal for finding the North-West Passage. Still enshrined in these two painted halls in the twenty-first century, longitude now appears in London as the preserve of major scientific and naval institutions, pursued and memorialised by great men. This book has shown what Hogarth's more ephemeral engraved version captured: that between these painted moments raged a paper war around the visual, mental, and social problems of longitude in the streets, taverns, and houses of London.

As the concerns of the Board of Longitude moved away from Harrison in London and more firmly towards exploration and observation at sea, the rake, longitude, and the network of visual, mental, and social concerns that they articulated went their separate ways in the works of Hogarth's successors. By 1835, the rake became the sailor in George Cruikshank's (1792–1878) *A Sailor's Progress* (Figure 81), concerned with the disparity between the round of niceties and appearance required of him at home,

13 Pressly, *James Barry's Murals*, particularly pp.91–112 on Commerce.

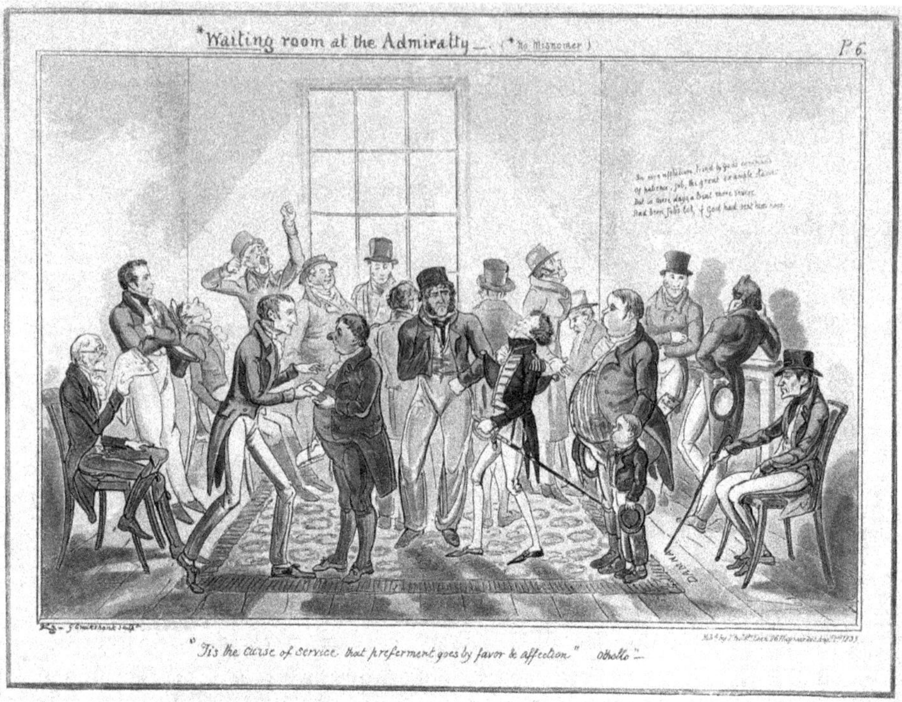

Figure 81: George Cruikshank, *A Sailor's Progress, Plate 6* (1835); coloured aquatint, 215 x 286mm, Royal Museums Greenwich, London.

and the grim and boisterous reality of naval life. Maps and diagrams became more central to an iconography of emerging national identity in Robert Dighton's 1793 *Geography Bewitched* (Figure 82), where the carefully mapped contours of the British Isles are a witty representation of John Bull on a fish, leaping out of a sea filled with naval vessels. Bedlam became a specific comment on political enthusiasm in James Gillray's (1756–1815) 1789 *Cooling the Brain* (Figure 83), where our lunatic has become politician Edmund Burke (1729–97), scrawling political threats on the wall of his cell (complete with Hoadly's original poem beneath). Longitude remained attached for longer to jokes around social boundaries and impolite behaviour. By the time Cruikshank's *A Comic Alphabet* appeared in 1836 (Figure 84) it had become a firm part of the vocabulary of social interaction. The absurdly dimensioned man and woman who denote the 'Latitude and Longitude' of society feature with such subjects as 'A la Mode' and 'Quadrille', but theirs was no longer the same longitude as that used by the sailors. The moment of longitude, captured by Hogarth, had passed.

Figure 82: Robert Dighton, *Geography Bewitched* (1793); hand-coloured etching, 209 x 173mm, © Trustees of the British Museum.

Figure 83: James Gillray, *Cooling the Brain* (1789); hand-coloured etching, 290 x 400mm, Yale University Library.

Figure 84: George Cruikshank, 'L [is for] Latitude and Longitude' from *A Comic Alphabet* (1836); facsimile coloured etching, 130 x 86mm, Private Collection.

Bibliography

including all known published pamphlets on longitude

Manuscript and Archival Collections

Beinecke Rare Book and Manuscript Library, Yale University, New Haven

Faccio, Nicholas, 'Letter: Worcester, England, to William Whiston, 1728 Oct 30', Gen MSS FILE 85

Bodleian Library, University of Oxford, Oxford

An Abstract of the Act of Parliament concerning the Discovery of the Longitude, and other Improvements of Navigation (London, 1714), Arch. A d.37

'A Short Ode upon ye Longitude', Rawlinson Manuscripts, MSS Rawl Poet 152, f.97

British Library, London

Samuel Clark to Hans Sloane, 15 November 1733, Sloane Collection, Sloane MS 4053 f.82

Caird Library, Royal Museums Greenwich, London

Barrington, William Wildman, 2nd Viscount Barrington, Barrington Papers, BGN/1–14

Logs of Lieutenants Thomson and Knowler, *H.M.S. Orford*, Admiralty Papers, ADM/L/O/29A

'Mr Whiston's Project for finding the Longitude' (6 November 1734), MSS/79/130.2

Wright, Robert, 'Viaticum Nautarum or The Sailor's Vade Mecum. Whereby they may be assisted to find both Latitude and Longitude by the help of certain fixed Stars collected out of Mr Flamsteed's Historia Coelestis, and of Tables truly Calculated from the great Sr. Isaac Newtons Complete Theory of the Moon', NVT/5

Cambridge University Library, Cambridge
'Guardbook of Letters', Papers of John Flamsteed, Royal Greenwich Observatory Archives, RGO 1/37
Newton, Isaac, 'Report to the Lords of the Admiralty on the Different Projects for Determining the Longitude at Sea', Newton Papers, MS Add.3972, http://cudl.lib.cam.ac.uk/view/MS-ADD-03972/63
Papers of the Board of Longitude, RGO 14/1–68, http://cudl.lib.cam.ac.uk/collections/longitude

Harrison Papers, Worshipful Company of Clockmakers, Guildhall Library, London
Description, with plans, by John Harrison, clockmaker, of his first longitude timekeeper or 'sea clock', CLC/L/CD/E/042/MS06026/001
Essays, notes and calculations, CLC/L/CD/E/037/MS03972/002
'An Explanation of my Watch or Timekeeper for the Longitude' and 'A Description of the Nature or Phenomenon of Ballances as found from Experience in my 3rd Machine etc', CLC/L/CD/E/036/MS03972/001
'Letters mainly from William Harrison, statements, cases etc relating to the tests at sea of John Harrison's longitude watches and his efforts to win an award from the Board of Longitude', CLC/L/CD/E/043/MS06026/002
Mechanical Notes c.1726–c.1772 and Drawings c.1730–72, CLC/L/CD/E/038/MS03972/003
Papers relating to the horological inventions of John Harrison, collected by Alexander Cumming, by virtue of his appointment by Act of Parliament to adjudicate on Harrison's explanation of the mechanisms of his longitude time keeper 'H4', CLC/L/CD/E/041/MS03973

National Archives, Kew
'Sentence of James Clarke of Calne, Wiltshire, 16 February 1721', Public Records Office, PROB 11/582/14

Private Collection
'Harrison Journal' MS

Royal Society Archives, London
'Description of Mr Whiston's Reflecting Telescope for the discovery of the Longitude at Sea' (1734), RBO 19/1
'James Logan, dated at Pennsylvania, to Edmond Halley, 25 May 1732', MS EL/L6/59
Laurans, 'Scheme of the Longitude or a piece of Watch worke of new Construction that shal goe upon Sea as regular as the best pendulum Clock shal goe upon Land', Cl.P/3ii/13

State Library of Victoria, Melbourne, Australia
'Account of John Harrison and his chronometer' (1817), MS Box 110/9 H17809

Printed Primary Works

Addison, John, *The Guardian* (2 vols, London, 1714)
Akenside, Mark, *The Pleasures of Imagination. A poem in three books* (London, 1744)
Algarotti, Francesco, *Sir Isaac Newton's philosophy explain'd for the use of the ladies. In six dialogues on light and colours* (London, 1739)
—— *Il Newtonianismo per le Dame: ovvero, dialoghi sopra la luce e i colori* (Naples, 1737)
Alimari, Dorotheo, *The new method of Signr Dorotheo Alimari Professor of the Mathematicks in Venice, for discovering the longitude* (London, 1714)
A. M., in Hydrostat, *Thoughts of a project for draining the Irish channel* (Dublin, 1722)
The Annual Register, or, A view of the history, politics, and literature for the year ... (London, 1762–1838)
'The Anti-Line of Beauty', *Public Advertiser* (15 February 1754) in Ronald Paulson, *Hogarth, His Life, Art and Times* (2 vols, New Haven, 1971), ii, p.493
A. R., *The Curiosities of Paris, in nine letters. Containing The Manner of Travelling from Calais to Paris, and the Description of the Towns. &c.* (London, c.1760)
Arbuthnot, John, *The Correspondence of Dr John Arbuthnot* (ed. Angus Ross) (Munich, 2006)
—— *Miscellaneous Works of the late Dr. Arbuthnot. With an account of the author's life. In two volumes* (London, 1770)
—— *Annus Mirabilis, or the Wonderful Effects of the approaching Conjunction of the Planets* Jupiter, Mars *and* Saturn, *by Abraham Gunter,* Philomath, *A Well-Wisher to the Mathematics* (London, 1722)
—— *To the Right Honourable the Mayor and aldermen of the City of London: The Humble Petition Of the Colliers, Cooks, Cook-Maids, Black-Smiths* (London, 1716)
—— *An Essay on the Usefulness of Mathematical Learning in a Letter from a Gentleman in the City to his Friend in Oxford* (London, 1701)
Atlas maritimus & commercialis; or, A general view of the World, so far as relates to trade and navigation (London, 1728)
Author of The Deluge and Bottomless Tub, *The longitude discover'd; A Tale* (London, 1726)
A. W., Esq., *The enormous abomination of the hoop-petticoat, as The Fashion Now is, And has been For about these Two Years Fully Display'd* (London, 1745)
Bailey, N., *An Universal Etymological English dictionary* (London, 1721)
Baker, Samuel, *A catalogue of upwards of ten thousand volumes of the best books in all arts and sciences* (London, 1754)
Bakewell, Thomas, after William Hogarth, Colophon Explanations to *A Rake's Progress* (1735)
Ballard, Thomas, *Bibliotheca Salmoneana, pars prima. Or, a catalogue of part of the library of the learned William Salmon, M.D. deceas'd* (London, 1713)
Bamfield, Samuel, *A new treatise of astronomy, or astronomy opened: being a rational and mathematic enquiry into the true principles and nature of astronomy* (Exeter, 1764)

The batchelor: or speculations of Jeoffry Wagstaffe, Esq (2 vols, Dublin, 1769)

Battie, William, *A Treatise on Madness. By William Battie M. D. Fellow of the College of Physicians in London, And Physician to St. Luke's Hospital* (London, 1758)

Beauty's triumph: or, The superiority of the fair sex invincibly proved (London, c.1745)

The Bee. Being essays on the most interesting subjects (London, 1759)

Ben Johnson's Jests; or, The wit's pocket companion (London, 1751)

Bernoulli, Jean, *Lettres Astronomiques où l'on donne une idée de l'état actuel de l'astronomie pratique dans plusieurs villes de l'Europe* (Berlin, 1771)

Billingsley, Case, *The longitude at sea, not to be found by firing guns, nor by the Most Curious Spring-Clocks or Watches* (London, 1714)

—— 'Since the publication of my late proposal, I find it is the opinion of many learned in astronomy and navigation, that the longitude at sea, will never be discovered to any degree of certainty' (London, 1714)

Bird, John, *The method of constructing mural quadrants* (London, 1768)

—— *The method of dividing astronomical instruments* (London, 1767)

Blackborrow, Peter, *The longitude not found, or, An answer to a treatise written by Henry Bond, Senior, shewing a way to find the longitude by the magnetical inclinatory needle wherein is proved that the longitude is not, nor cannot be found by the magnetical inclinatory needle* (London, 1678)

Blackmore, Richard, Sir, *The Lay-Monastery. Consisting of essays, discourses, &c.* (London, 1714)

Blundel, William, *To all and Singular the Honourable Commissioners, constituted by Act of Parliament, for the discovery of the longitude at sea* (Coventry, 1721–23)

—— 'Advertisement. To all those in general that are wellwishers to the mathematicks, and have a desire to promote the honour, good, and benefit, of Great Britain' (Coventry, c.1721)

Bond, Henry, *The longitude found, or, A treatise shewing an easie and speedy way, as well by night as by day, to find the longitude, having but the latitude of the place and the inclination of the magnetical inclinatorie needle* (London, 1676)

Breval, John, *Play is the Plot. A comedy* (London, 1718)

Brown, Thomas, *The Third Volume of the Works of Mr. Thomas Brown, containing, Amusements serious & comical, calculated for the meridian of London* (London, 1708)

Browne, John, *A defence of the Bishop of London's second pastoral letter, against the exceptions advanc'd in a late piece* (London, 1730)

Browne, Robert, *To the Honourable the Commons of Great Britain in Parliament assembled, the case of Robert Browne humbly sheweth* (London, 1732)

—— *A new improvement of the theory of the moon* (London, 1731)

—— *The Discovery of the Longitude at Sea by Celestial Observations Ascertained and humbly offered* (London, 1730)

—— *To the Right Honourable, and Honourable, the Commissioners, appointed by Act of Parliament, for providing a Publick Reward for such Person, or Persons, that shall find the Longitude at Sea. The humble Petition of Robert Browne* (London, 1729)

—— *Methods, propositions and problems, for finding the latitude; With the Degree and Minute of the Equator upon the Meridian. And The Longitude at Sea. By Caelestial Observations only. And also by Watches, Clocks, &c. and to correct them and know their Alterations* (London, 1714)

Bull, Digby, *A letter of advice to all the worthy and ingenuous merchants of the city of London, and elsewhere in England, Scotland, and Ireland; And also to all the Worthy and Ingenuous Merchants of France, and of Holland and the United-Provinces, and of Germany, Spain, Portugal, Denmark, and Sweden, and elsewhere in Europe; and To all Ingenuous and Ingenious Artists in Astronomy, Geography, and Navigation; shewing An exact, easie and speedy Way to know the Longitude of all Places in the World, where the European Merchants have their Agents to make Observations; and also How the Longitude of Places may be better known upon Ship-Board* (London, 1706)

Burgess, Daniel, *A letter to the Bishop of Salisbury, occasion'd by his son's letter to the Earl of Hallifax* (London, 1715)

Burgh, James, *Crito, or, essays on various subjects* (2 vols, London c.1766–67), i

Butler, Samuel, *Hudibras by Samuel Butler Tom. I. I* (3 vols, London, 1793), i

A Calculation shewing the result of an Experiment made with Mr Harrison's Time-Keeper, in Consequence of the Instructions and Directions of the Commissioners of the Longitude, in a Voyage from Portsmouth to Port Royal in Jamaica, in Order to Determine by that Time-Keeper, the Difference of Longitude between these two Places; and to find how much the Longitude found by the Time-Keeper, differs from the Longitude formerly determined by Astronomical Observations (London, 1762)

Carter, Elizabeth, *A series of letters between Mrs. Elizabeth Carter and Miss Catherine Talbot, from the year 1741 to 1770. To which are added, Letters from Mrs. Elizabeth Carter to Mrs. Vesey, between the years 1763 and 1787, published from the original manuscripts in the possession of the Rev. Montagu Pennington, M.A. vicar of Northbourn in Kent, her nephew and executor* (4 vols, London, 1809), i

A catalogue of all the English books that have been published for these sixty years past (London, 1764)

Cater, William, *A catalogue of many thousand valuable and elegant books, in fine condition, well preserv'd and richly bound* (London, 1770)

Cawthorn, James, *Poems, by the Rev. Mr. Cawthorn. Late master of Tunbridge School* (London, 1771)

Chambers, Ephraim, *Cyclopædia: or, an universal dictionary of arts and sciences* (London, 1728)

Cheyne, George, *The English Malady* (ed. Roy Porter) (London, 1991)

'Chloe's Play-Thing. A Tale' in *The muse in good humour: or, a collection of comic tales, &c. from Dryden, Congreve, Gay, Sedley, Cobb, and other eminent poets* (London, 1745), pp.126–30

Clarke, James, *The mercurial chronometer improv'd: or, a supplement to a book entituled, An essay, wherein a method is humbly propos'd for measuring equal time with the utmost exactness* (London, 1715)

—— *An essay, wherein a method is humbly propos'd for measuring equal time with the utmost exactness* (London, 1714)

The Clockmakers Outcry against the Author of the Life and Opinions of Tristram Shandy (London, 1760)

Cocks, Richard, Sir, *A perfect discovery of the longitude. By Sir Richard Cocks, Bart. Dedicated to the Earl of Peterborough* (London, 1721)

Cole, Benjamin, *The description and use, of a new quadrant, for finding the latitude at sea* (London, 1748)

A Collection of the Occasional Papers for the Year 1716. With a preface (London, 1716)

Conolly, Mr., *The Connoisseur: or, every man in his folly. A comedy* (London, 1736)

Cooper, Mrs., *The Rival Widows: or, fair libertine. A comedy* (London, 1735)

Costard, George, *The history of astronomy, with its application to geography, history, and chronology; occasionally exemplified by the globes* (London, 1767)

Coster, John, *A Practicable Method for finding the Longitude at Sea, by a Marmeter, or Instrument, for Measuring the exact Run of the Ship* (London, 1714)

Crathorne, Jonah, *The longitude demonstrated, by the use of a silent pendulum; that keeps true time at sea* (London, 1718)

Curious enquiries, Being six brief discourses, viz. I. Of the longitude. II. The tricks of astrological quacks. III. Of the depth of the sea. IV. Of tobacco. V. Of Europes being too full of people. VI. The various opinions concerning the time of keeping the Sabbath (London, 1688)

The Daily Courant (London, 1702–35)

The Daily Journal (London, 1721–37)

The Daily Post (London, 1719–46)

A Declaration of the earthly flatform (London, 1679)

Defoe, Daniel, *An Essay upon Projects* (London, 1697)

Delany, Patrick, *Observations upon Lord Orrery's Remarks on the life and writings of Dr. Jonathan Swift* (London, 1754)

Derham, William, *The Artificial Clock-Maker. A treatise of watch and clock-work, shewing to the meanest capacities the art of calculating numbers to all sorts of movements* (London, 1714)

—— *Physico-Theology: or, a demonstration of the being and attributes of God, from his works of creation. Being the substance of XVI sermons Preached in St. Mary le Bow-Church, London, at the Honble Mr. Boyle's Lectures, in the Years 1711 and 1712* (London, 1714)

De Saumarez, Henry, *A letter to a friend, setting forth the case of the author of the marine surveyor. Being an instrument design'd for the improvement of navigation, and for correcting many errors in that useful art, on which the happiness of Great Britain so much depends* (London, 1725)

—— *An account of the proceedings of Henry De Saumarez, of the island of Guernsey, gent. Concerning his discovery of an invention, by which the course of a ship at sea may be better ascertained than by the logg-line* (London, 1717)

The Description and Uses of Barston's New-Invented Universal Astronomical Quadrant (London, 1740)

Dollond, P. and J., *Directions for using the new improved Hadley's quadrant; with the new method of adjusting the glasses for the back observation* (London, c.1772)

Dunn, Samuel, *The navigator's guide, to The Oriental or Indian Seas: or, the description and use of a variation chart of the magnetic needle* (London, 1775)
—— *A new and general introduction to practical astronomy: with its application to geography, in describing the Earth* (London, 1774)
Dunton, John, *The conventicle: or, a narrative of the dissenters new plot against the present constitution in church and state* (London, 1714)
—— *The art of living incognito being a thousand letters on as many uncommon subjects / written by John Dunton during his retreat from the world, and sent to that honourable lady to whom he address'd his conversation in Ireland; with her answer to each letter* (London, 1700)
—— *A voyage round the world, or, A pocket-library divided into several volumes* (London, 1691)
Ellicott, John, *A description of two methods, by which the irregularities in the motion of a clock, arising from the influence of heat and cold upon the rod of the pendulum, may be prevented* (London, 1753)
Encyclopaedia Britannica; or, a dictionary of arts and sciences, Compiled upon a New Plan (3 vols, Edinburgh, 1771)
An essay towards a new method to shew the longitude at sea; especially near the dangerous shores (London, 1714)
Euler, Leonhard, *Lettres a une princesse d'allemagne sur divers sujets de physique et de philosophie* (3 vols, The Hague, 1775)
Euler, Leonhard, and Mayer, Tobias, *The Euler–Mayer Correspondence (1751–1755)* (ed. Eric Forbes) (New York, 1971)
Explanation of the principles and use of a new constructed quadrant (London, c.1770)
Fatio de Duillier, Nicolas, *Navigation improv'd: being chiefly the method for finding the latitude, at sea as well as by land, by taking any proper altitudes, together with the time between the observations* (London, 1728)
Ferguson, James, *Syllabus of a course of lectures on the most interesting parts of mechanics, hydrostatics, hydraulics, pneumatics, electricity, and astronomy* (Edinburgh, 1768)
—— *The young gentleman and lady's astronomy, familiarly explained in ten dialogues between Neander and Eudosia* (Dublin, 1768)
Ferrner, Bengt, *Resa i Europa. En astronom, industrispion och teaterhabitué genom Danmark, Tyskland, Holland, England, Frankrike och Italien 1758–1762* (ed. Sten G. Lindberg), *Studies and Sources Published by the Swedish History of Science Society 14* (Uppsala, 1956)
Fielding, Henry, *The Masquerade, A Poem. Inscribed to C—t H—d—g—r. By Lemuel Gulliver, Poet Laureate to the King of Lilliput* (London, 1728)
Fitzgerald, Thomas, 'Bedlam' in *Poems on several occasions* (London, 1736), pp.1–11
Flamsteed, John, *The Correspondence of John Flamsteed, The First Astronomer Royal, Volume Three, 1703–1719* (ed. Eric G. Forbes) (Bristol, 2001)
Flavel, John, *Navigation spiritualiz'd: or, A new compass for seamen* (London, 1698)
Fontenelle, M. de, *Entretiens sur la Pluralité des Mondes* (The Hague, 1728)
—— *Entretiens sur la Pluralité des Mondes* (Paris, 1687)

Formey, Mr., 'The life of John Philip Baratier, the prodigy of this age for genius and learning' in *The History of the Works of the Learned for October 1743* (London, 1745)

The Fortune-Hunter: or, the Gamester Reclaim'd. A play. With an entertainment of the millenium. Representing Paradise lost: or, the fall of man (London, c.1775)

Franklin, Benjamin, *Memoirs of the Life and Writings of Benjamin Franklin, LL.D., F.R.S., &c.* (ed. William Temple Franklin) (3 vols., London, 1817–18), http://franklinpapers.org/franklin/framedVolumes.jsp

French, John, *A perfect discovery of the longitude at sea. In compliance with what's propos'd in a late act of Parliament* (London, 1715)

Fyler, Samuel, *Longitudinis inventae explicatio non longa, or, Fixing the volatilis'd, and taking time on tiptoe, briefly explain'd by which rules are given to find the longitude at sea by, as truly and exactly as the latitude is found by the star in the tayle of Ursa Minor, call'd the Pole-star* (London, 1699)

Gay, John, 'A True and faithful Narrative of what pass'd in London during the general Consternation of all Ranks and Degrees of Mankind; On Tuesday, Wednesday, Thursday, and Friday last' in Vinton A. Dearing (ed.), *John Gay: Poetry and Prose* (2 vols, Oxford, 1974), ii, pp.464–73

Gentleman, *Man superior to woman; or, a vindication of Man's Natural Right of Sovereign Authority over the Woman. Containing a plain Confutation of the fallacious Arguments of Sophia, in her late Treatise intitled, Woman not inferior to Man* (London, 1739)

Gentleman on His Travels Thro' Germany, *The German Spy: or, familiar letters from a gentleman on his travels thro' Germany, to his friend in England* (London, 1740)

The Gentleman's Magazine (London, 1736–1833)

Gentleman who was concern'd, *Considerations on the unhappy accidents that attended the last undertaker in his works at Dagenham Breach* (London, 1723)

Gilpin, William, *An Essay upon Prints: containing remarks upon the principles of picturesque beauty; the different kinds of prints; and the characters of the most noted masters* (London, 1768)

Gordon, John, *Occasional thoughts on the study and character of classical authors, on the course of litterature, and the present plan of a learned education* (London, 1762)

Griffith, Richard, *The Triumvirate: or, the authentic memoirs of A. B. and C* (2 vols, London, 1764), i

The Grub Street Journal (London, 1730–37)

Hadley, John, 'The Description and Use of a New Instrument for taking Angles, by Reflection, invented by Mr. J. Hadley' in *The Philosophical Transactions (from the Year 1719, to 1733) Abridged, and Disposed under General Heads* (ed. John Eames and John Martyn) (2 vols, London, 1734), pp.139–45

—— *A description of a new instrument for taking the latitude or other altitudes at sea. With directions for its use* (London, 1734)

—— 'The Description and Use of a New Instrument for Taking Angles' in *Philosophical Transactions* 37 (1731–32), pp.148–50

Haldanby, Francis, *An attempt to discover the longitude at sea, pursuant to what is proposed in a late act of Parliament* (London, 1714)

Hales, Stephen, *A description of ventilators: whereby great quantities of fresh air may with ease be conveyed into mines, goals, hospitals, Work-Houses and Ships, In Exchange for their Noxious Air* (London, 1743)

Halley, Edmond, *Correspondence and Papers of Edmond Halley, preceded by an unpublished memoir of his life by one of his contemporaries and the* 'éloge' *by D'Ortous de Mairan* (ed. Eugene Fairfield MacPike) (Oxford, 1932)

Hamilton Moore, John, *The Practical Navigator and Seaman's new Daily Assistant. Being a complete system of practical navigation* (London, 1772)

Hanna, John, *An Astronomical Creed, in ten articles, with natural and moral reflections from them; to which is subjoined, the new method of discovering the longitude, by the appulse of the moon to the fixed stars: improved by Sir Isaac Newton's theory of the moon and Mr. Flamsteed's Geometrical construction of the parallax* (Dublin, 1725)

Hanway, Jonas, *Thoughts on the importance of the Sabbath, with a caution not to trespass on the design of it* (London, 1765)

Hardy, Samuel, *The Theory of the Moon made Perfect: So far at least as to determine the Longitude both at Sea and Land, within the limits required by Act of Parliament. To which is added an Account of the Use which may be made of Comets* (London, 1751)

Harris, John, *Astronomical Dialogues between a gentleman and a lady: wherein the doctrine of the sphere, uses of the globes, and the elements of astronomy and geography are explain'd* (London, 1729)

—— *Lexicon Technicum: or, An Universal English Dictionary of Arts and Sciences: Explaining not only the terms of art, but the arts themselves* (2 vols, London, 1723)

—— *Lexicon Technicum: or, An Universal English Dictionary of Arts and Sciences: Explaining not only the terms of art, but the arts themselves* (London, 1704)

Harrison, Edward, *Idea Longitudinis; being a brief definition of the best known axioms for finding the longitude, or a more rational discovery thereof, than hath been heretofore published* (London, 1696)

Harrison, John, *A description concerning such mechanism as will afford a nice, or True mensuration of time; Together with some account of the attempts for the discovery of the longitude by the moon: as also an account of the discovery of the scale of musick* (London, 1775)

—— *The Case of Mr. John Harrison* (London, c.1770)

—— *The Case of Mr. John Harrison* (London, c.1767)

—— *The Principles of Mr. Harrison's Time-Keeper, with plates of the same. Published by order of the Commissioners of Longitude* (London, 1767)

—— *Remarks on a Pamphlet lately published by the Rev. Mr Maskelyne, under the Authority of the Board of Longitude* (London, 1767)

—— *A narrative of the proceedings relative to the discovery of the longitude at sea; by Mr. John Harrison's time-keeper; Subsequent to those published in the Year 1763* (London, 1765)

—— *An account of the proceedings, in order to the discovery of the longitude: In a letter to the Right Honourable ******, Member of Parliament* (London, 1763)

Haselden, Thomas, *The seaman's daily assistant, being a short, easy, and plain method of keeping a journal at sea; in which are contain'd rules, shewing how the allowances for lee-way, Variation, Heave of the Sea, Set of Currents, &c. are to be made* (London, 1761)

—— *The description and use of that most excellent invention commonly call'd mercator's chart* (London, 1722)

—— *A reply to Mr. Wilson's answer to my letter to Dr. Halley* (London, 1722)

Hatton, Thomas, *Introduction to the mechanical part of clock and watch work. In two parts* (London, 1773)

Hautefeuille, Jean de, *Machine loxodromique, qui trace sur un papier, en telle proportion que l'on veut, le chemin que fait un navire, par le moyen de laquelle les pilotes auront facilement la connoissances des longitudes* (Paris, 1750)

Hawkins, Isaac, *An essay for the discovery of the longitude at sea, by several new methods fully and particularly laid before the publick* (London, 1714)

Haywood, Eliza Fowler, *The Husband. In answer to The wife* (London, 1756)

Heath, Robert, *The British Palladium: or annual miscellany of literature and science: for the year 1775* (London, 1775)

—— *The British Palladium: or, annual miscellany of literature and science: for the year 1773* (London, 1773)

—— *The Seaman's Guide to the Longitude: or a Key to the Nautical Almanac and Astronomical Ephemeris* (London, 1770)

—— *The British Palladium; or, Annual miscellany: for the bissextile year 1768* (London, 1768)

—— *A Supplement to the Royal Astronomer and Navigator. Supplying (by reduction and equation) accurate astronomical tables* (London, 1768)

—— *The Palladium-Supplement, enlarged. (price 1s.) For the year of our Lord, 1764* (London, 1764)

—— *A New and Easy Practical Method of determining the Longitude, at Sea or Land, to Half a Degree, from the Royal Observatory at Greenwich. Also, Remarks on the speculative and fallible Methods that have been proposed for discovering the Longitude* (London, 1761)

—— *A natural and historical account of the islands of Scilly; describing their situation, number, extent, Soil, Culture, Produce, Rarities, Towns, Fortifications, Trade, Manufacture, Inhabitants* (London, 1750)

Hebert, Jean, *Traité des Longitudes, ou Navigation dans son jour, Où est montré la maniere de connoître les Longitudes, tant sur Mer que sur Terre, d'une méthode nouvelle & facile à pratiquer* (Paris, 1718)

Hill, John, *The Inspector* (2 vols, London, 1753)

Historical Manuscripts Commission, *Report on The Manuscripts of his Grace the Duke of Portland Preserved at Welbeck Abbey* (Norwich, 1899), Volume V

Hobbs, William, *A new discovery for finding the longitude. Humbly submitted to the approbation of the right honourable the lords spiritual and temporal, and the rest of the honourable persons, appointed by the late act of Parliament, for hearing and determining proposals relating to the said longitude* (1716)

——— *A new discovery for finding the longitude. Humbly submitted to the approbation of the right honourable the lords spiritual and temporal, and the rest of the honourable persons, appointed by the late act of Parliament, for hearing and determining proposals relating to the said longitude* (London, 1714)

Hogarth, William, *The Analysis of Beauty* (ed. Ronald Paulson) (Yale, 1997)

——— *The Case of Designers, Engravers, Etchers, &c. stated. In a Letter to a Member of Parliament* (New York, 1975)

——— 'Autobiographical Notes' in J. Burke (ed.), *The Analysis of Beauty, with the Rejected Passages from the Manuscript Drafts and Autobiographical Notes* (Oxford, 1955), pp.201–31

——— *The genuine works of William Hogarth: illustrated with biographical anecdotes, a chronological catalogue, and commentary / by John Nichols ... and the late George Steevens* (3 vols, London, 1808–17)

Hunt, William, *The projectors. A comedy* (Edinburgh, 1737)

Irwin, Christopher, *A summary of the principles and scope of a method, humbly proposed, for finding the longitude at sea* (London, 1760)

Jackson, Benjamin Habakkuk, *Some new thoughts founded upon new principles, concerning a threefold motion of the earth. The rectification of the Kallender. The flowing and ebbing of the sea. The nature of the magnet. The variation of the compass. The cause of sea currents, and trade winds. The various motions of all the coelestial orbs. The finding out the true place of the moon. And facilitating the discovery of the longitude* (London, 1714)

Jackson, Henry, *Longitude and latitude in conjunction, by sea and land, by day and night, found out and demonstrated, in a new way of astronomical observation* (London, 1727)

Jacob, Hildebrand, 'Bedlam' in *The works of Hildebrand Jacob, Esq; containing poems on various subjects, and occasions* (London, 1735), pp.11–25

Jemmy Carson's Collections: being a revival of his own labours and lucubrations, for thirty years past (Dublin, 1744)

Jenner, Charles, *Letters from Altamont in the capital, to his friends in the country* (London, 1767)

John, Gabriel, *Flagellum: or, a dry answer to Dr. Hancock's wonderfully-comical liquid book, which he merrily calls Febrifugum magnum* (London, 1723)

Johnson, Samuel, *A Dictionary of the English language: In Which the Words are Deduced from their Originals, and Illustrated in their Different Significations* (2 vols, London, 1755–56)

Journals of the House of Commons 1688–1834, House of Commons Parliamentary Papers online, http://parlipapers.chadwyck.co.uk/home.do

Keach, Benjamin, *The Progress of Sin; or, the travels of ungodliness. ... In an apt and pleasant allegory* (London, 1763)

Keith, George, *An Easie Method not to be found hitherto in any Author, or History, whereby the Longitude of any Places at Sea or Land, West or East from any first Meridian, E.g.* London, *may be found at any Distance, great or small, by certain fixed Stars* (London, 1713)

Kenrick, W., *An address to the artists and manufacturers of Great Britain; Respecting an Application to Parliament for the farther Encouragement of New Discoveries and Inventions in the Useful Arts* (London, 1774)

Kersey, John, *A New English Dictionary: Or, Compleat collection of the most proper and significant words, commonly used in the language; with a short and clear definition of difficult words and terms of Art* (London, 1702)

Kirby-Miller, Charles (ed.), *Memoirs of the Extraordinary Life, Works and Discoveries of Martinus Scriblerus, written in collaboration by the Members of the Scriblerus Club: John Arbuthnot, Alexander Pope, Jonathan Swift, John Gay, Thomas Parnell, and Robert Harley, Earl of Oxford* (Oxford, 1988)

The Ladies Diary: or, the Womens Almanack (London, 1706–1841)

Lady of Quality, *Woman triumphant: or, the excellency of the female sex; asserted in opposition to the male* (1721)

Lalande, Jérôme, *Diary of a Trip to England 1763* (ed. and transl. Richard Watkins) (Kingston, Tasmania, 2002)

Leadbetter, Charles, *Uranoscopia: or, the contemplation of the heavens. Being a demonstration of the equation of time. With the method of observing the solar ingresses into any point of the ecliptic* (London, 1735)

The Life and Opinions of Tristram Shandy, Gentleman. Vol. IX (London, 1766)

Lisle de la Drevetière, Louis François de, *La Decouverte des Longitudes avec La Méthode facile aux Navigateurs pour en faire usage actuellement* (Paris, 1740)

Locke, John, *An Essay Concerning Human Understanding. In four books* (London, 1706)

Locke, Richard, *The new and universal problem to discover the longitude at sea: further explained and illustrated* (London, 1752)

—— *A new problem to discover the longitude at sea, by the same observation, and with the same certainty, as the latitude is: together with the translation, of the problem into Latin* (London, 1751)

—— *The Longitude* (London, 1732)

—— *The circle squared. To which is added, a problem to Discover the longitude both at land and sea, by Means of a New Instrument which Discovers the Hour of the Day or Night in any Latitude, as well as the Latitude it self* (London, 1730)

The London and Westminster Guide, Through the Cities and Suburbs (London, 1768)

The London Spy. By the Author of the Trip to Jamaica (London, 1698–1700)

Ludlam, William, *Directions for the use of Hadley's quadrant, with remarks on the construction of that instrument* (London, 1771)

Lund, John, *The mirrour: A Poem. In imitation of C. Churchill. To which are added, three tales, (in the manner of Prior)* (London, 1771)

Lyttelton, George, *Letters from a Persian in England, To his Friend at Ispahan* (London, 1735)

Malassis de Sulamar, John-Baptist, *Perpetual universal longitude discovered on Monday the 30th of April of the last year bissextile 1764* (London, 1765)

Martial, *Epigrams of Martial, &c. with mottos from Horace, &c. Translated, imitated, adapted, and addrest to the Nobility, Clergy, and Gentry* (London, 1773)

Martin, Benjamin, *The Young Gentleman and Lady's Philosophy, in a Continued Survey of the Works of Nature and Art; by way of a Dialogue* (London, 1772)

—— *Institutions of Astronomical Calculations* (London, 1765)

—— 'Biographia Philosophica' in *The General Magazine of Arts and Sciences, philosophical, philological, mathematical, and mechanical* (14 vols, London, 1755–65), xiii

—— 'The Young Gentleman and Lady's Philosophy' in *The General Magazine of Arts and Sciences, philosophical, philological, mathematical, and mechanical* (14 vols, London, 1755–65), i

—— *A Course of Lectures in Natural and Experimental Philosophy, Geography and Astronomy* (Reading, 1743)

Maskelyne, Nevil, *Remarks on the Hadley's quadrant, tending principally to remove the Difficulties which have hitherto attended the Use of the Back-Observation, and to obviate the errors that might arise from a want of parallelism in the two surfaces of the index-glass* (London, 1772)

—— *An account of the going of Mr. John Harrison's watch, at the Royal Observatory, from May 6th, 1766, to March 4th, 1767. Together with the original observation and calculations of the same* (London, 1767)

—— *The British Mariner's Guide. Containing, complete and easy instructions for the discovery of the longitude at sea and land, within a Degree, by Observations of the Distance of the Moon from the Sun and Stars, taken with Hadley's Quadrant* (London, 1763)

Mayer, Tobias, *Tabulæ motuum solis et lunæ novæ et correctæ; auctore Tobia Mayer: quibus accedit Methodus longitudinum promota, eodem auctore. Editæ jussu Præfectorum Rei Longitudinariæ* (London, 1770)

A Method for finding, wherever we are, by an observation of the Moon, more practicable at Sea than at Land, What o'clock it then is at London; and consequently for finding the Longitude of the Place where our Ship may then happen to be (London, 1763)

Miller, James, *Of Politeness. An epistle to the Right Honourable William Stanhope, Lord Harrington. The second edition* (London, 1738)

—— *The Humours of Oxford. A comedy* (London, 1730)

Montfichet, Bertram, *The Life and Opinions of Bertram Montfichet, Esq; written by himself* (London, c.1761)

The Monthly review; or, Literary journal. Giving a candid account, with abstracts of, or extracts from, the new books and pamphlets, published in Great-Britain and Ireland, as they come out. … By several hands (London, 1752–1789)

Morning Post and Daily Advertiser (London, c.1773–1792)

Mountaine, William, *An account of the methods used to describe lines, on Dr. Halley's chart of the terraqueous globe, shewing the variation of the magnetic needle about the year 1756* (London, 1758)

—— *An account of the methods used to describe lines, on Dr. Halley's chart of the terraqueous globe, shewing the variation of the magnetic needle about the year 1744* (London, 1746)

Murray, Hamilton, *The Life and Real Adventures of Hamilton Murray. Written by himself* (3 vols, London, 1759), ii

Navigation Improved: or, some new and easy inventions to find a ship's true course always, a ship's latitude at any hour; and the longitude at sea: never before disclosed (London, 1741)

Newton, Isaac, Sir, *The Correspondence of Isaac Newton* (ed. H. W. Turnbull) (7 vols, Cambridge, 1959–77)

Newton, Samuel, *The Art of Navigation, containing Easy Rules for finding the Latitude and Difference of Longitude of Places by observation of the sun, moon and stars* (London, 1715)

Nichols, J., *Biographical Anecdotes of William Hogarth; and a Catalogue of his Works chronologically arranged; with Occasional Remarks* (London, 1781)

N. N., *Will-With-a-Wisp; or, the grand ignis fatuus of London. Being a lay-man's letter to a country-gentleman, concerning the articles lately exhibited against Mr. Whiston* (London, 1714)

'Ode, for Musick, on the Longitude' in *Miscellanies. The Last Volume* (London, 1727)

Parker, Benjamin, *A survey of the six days works of the creation: Philosophically proving the Truth of the Account thereof, as deliver'd by Moses in the First Chapter of Genesis* (London, 1745)

—— *Parker's projection of the longitude at sea: whereby the mariners may be enabled to correct their accounts thereof, as often as they shall have the benefit of a clear sky and calm sea, at the time of the moon's visible southing* (Nottingham, 1731)

Parliamentary Statutes online, http://www.justis.com/search.aspx?search=legislation

Partridge and Flamsted, *Dr. Flamstead's and Mr. Patridge's new fortune-book* (London, c.1730)

Philo-Deus, *Considerations on the Latitude of Sir Richard Cocks's New Opinions concerning the Longitude of the Church of England; wherein the most false and virulent Aspersions on the Clergy are detected, and the doughty Champion for Irreligion and Impiety call'd upon to a 2d Charge, with more Truth, and less Scandal, against that Body of Men: Recommended to his Curate of Dumbleton* (London, 1722)

The Philosophical Transactions (London, 1665–), http://rstl.royalsocietypublishing.org

Plank, Stephen, *An introduction to a true method for the discovery of longitude at sea. Humbly offer'd to the consideration of both Houses of Parliament* (London, 1720)

—— *An introduction to the only method for discovering longitude. Humbly presented to both Houses of Parliament, and to those Worthy Persons that are to judge of it* (London, 1714)

Pledger, Elias, *A Brief Description of a new Invented, Small, yet Accurate Astronomical Quadrant. Particularly calculated for Curious Observations at Sea, but equally useful at Land: And that either by Day or by Night* (London, 1731)

Pope, Alexander, *Selected Poetry* (ed. Pat Rogers) (Oxford, 1998)

The Post Boy (London, 1695–1728)

Post Man, and the Historical Account (London 1695–1730)

The Present State of the Republick of Letters (18 vols, London, 1728–36)

Principes de la montre de Mr. Harrison, avec les planches relatives a la même montre, imprimés à Londres en 1767 par ordre de Mrs. les Commissaires des Longitudes (Avignon, 1767)

Public Ledger or The Daily Register of Commerce and Intelligence (London, 1760–61)

Purshall, Conyers, 'To the Lords Commissioners for the Longitude' (London, c.1714)

——— *An essay at the mechanism of the macrocosm: or the dependance of effects upon their causes. In a new hypothesis, Accommodated to Our Modern and Experimental Philosophy* (London, 1705)

The Question concerning Literary Property, determined by the Court of King's Bench on 20th April, 1769, in the cause between Andrew Millar and Robert Taylor (London, 1773)

R*. R*., *An original camera obscura; or, The court, city, and country magic-lanthorn* (London, 1768)

Ralph, James, *The Fashionable Lady; or Harlequin's opera. In the Manner of a rehearsal* (London, 1730)

Ramsden, Jesse, *Description of an Engine for dividing Mathematical Instruments. By Mr. J. Ramsden, mathematical instrument-maker. Published by order of the Commissioners of Longitude* (London, 1777)

The Real seeker: Or, A series of letters relative to his religious doubts, with their respective solutions and rejoinders (London, 1767)

Robertson, J., *The Elements of Navigation; containing the theory and practice. With the necessary tables* (2 vols, London, 1772)

——— *The Elements of Navigation; containing the theory and practice. With the necessary tables* (2 vols, London, 1764)

Robins, Benjamin, *New principles of gunnery: containing, the determination of the force of gun-powder, and an investigation of the difference in the resisting power of the air to swift and slow motions* (London, 1742)

Rolt, Richard, *A New Dictionary of Trade and Commerce, compiled from the information of the most eminent merchants, and from the works of the best writers on commercial subjects, in all languages* (London, 1756)

Rowe, Jacob, *Navigation improved: In two books* (London, 1725)

Saavedra, Miguel de Cervantes, *The history of the renowned Don Quixote de la Mancha Written in Spanish by Miguel de Cervantes Saavedra. Translated by Several Hands* (4 vols, London, 1733)

Saavedra Fajardo, Diego de, *Respublica Literaria: or, the republick of letters; being a vision. Wrote in Spanish by Don Diego de Saavedra, Knight of the Order of St. James. Translated from the original by J. E. A. B.* (London, 1727)

——— 'Republica Literaria' in *Obras de Don Diego de Saavedra* (2 vols, Amberes, 1681)

A sailor's proposal for finding his longitude by the moon. Offering at an easy and practical way of performing it now, by the moon's visible apparent equinoctial distance. The whole Process being within the Power and Capacity of every Mariner who understands the Art of Navigation (London, 1726)

Sedgewick, O., *The Universal Masquerade: or, the world turn'd inside-out. Delineating and detecting the virtues and vices of mankind, From The Court to the Cottage, In all Professions, for the benefit of both sexes* (2 vols, London, 1742), ii

Smart, Christopher, *The Poetical Works of Christopher Smart, Volume I: Jubilate Agno* (ed. Karina Williamson) (Oxford, 1980)

Society of Gentlemen, *A New and Complete Dictionary of Arts and Sciences; Comprehending all the branches of useful knowledge, with Accurate Descriptions as well of the various Machines, Instruments, Tools, Figures, and Schemes necessary for illustrating them, as of The Classes, Kinds, Preparations, and Uses of Natural Productions, whether Animals, Vegetables, Minerals, Fossils, or Fluids; Together with The Kingdoms, Provinces, Cities, Towns, and other remarkable Places throughout the World* (4 vols, London, 1763)

Squire, Jane, *A proposal for discovering our longitude / Proposition pour la decouverte de notre longitude* (London, 1742)

The Star-gazer: or Latitude for Longitude, A Poem (1739, London)

Stayley, George, *The life and opinions of an actor. A real history, in two real volumes* (2 vols, Dublin, 1762)

Sterne, Laurence, *The Life and Opinions of Tristram Shandy, Gentleman* (ed. Graham Petrie) (London, 1985)

Stevens, John, *The modern wife; or, the virgin her own rival. A comedy* (London, 1744)

Stone, Edmund, *Some reflections on the uncertainty of many astronomical and geographical positions, with regard to the figure and magnitude of the earth, the finding the longitude at sea by watches, and other assertions of the most eminent astronomers. With some Hints towards their Reformation and Emendation* (London, 1766)

Stukeley, William, *The Family Memoirs of the Rev. William Stukeley, M.D. and the Antiquarian and other Correspondence of William Stukeley, Roger and Samuel Gale, etc.* (3 vols, Durham, 1883)

Sturm, Leonard Christofle, *Project de la Resolution du fameux problème touchant La Longitude sur Mer* (Nuremberg, 1720)

A supplement to Dean Sw—t's miscellanies: by the author. Containing, I. A letter to the students of both universities, relating to the New Discoveries in Religion and the Sciences, and the Principal Inventors of them. II. An essay upon an apothecary. III. An account of a surprizing apparition, October 20. 1722 (London, 1723)

Swift, Jonathan, *Gulliver's Travels* (ed. Claude Rawson) (Oxford, 2005)

—— *Major Works* (ed. Angus Ross and David Woolley) (Oxford, 2003)

—— *The Correspondence of Jonathan Swift* (ed. Harold Williams) (5 vols, Oxford, 1963–65)

—— *Journal to Stella* (ed. Harold Williams) (2 vols, Oxford, 1948)

—— 'Some Thoughts on Free Thinking written in England but left unfinished' in *An appendix to Dr. Swift's works and literary Correspondence. Improved From an Edition printing by Mr. Faulkner* (London, 1767), pp.14–16

—— *Some observations upon a paper, call'd, The report of the Committee of the most honourable the Privy-Council in England, relating to Wood's half-pence* (Dublin, 1724)

—— *The Bubble: A Poem* (London, 1721)

Thacker, Jeremy, *The longitudes examin'd. Beginning with a short epistle to the longitudinarians, and ending with the description of a smart, pretty machine of my own, which I am (almost) sure will do for the longitude, and procure me the twenty thousand pounds* (London, 1714)

Theaker, Robert. *A light to the longitude, or, The use of an instrument called the seaman's director speedily resolving all astronomical cases and questions concerning the sun, moon, and stars: with several propositions whereby sea-men may find at what meridian and longitude they are at in all parts of the world* (London, 1665)

Uffenbach, Zacharias Konrad von, *London in 1710: from the travels of Zacharias Conrad von Uffenbach* (transl. and ed. W. H. Quarrell and Margaret Mare) (London, 1934)

The Universal Magazine of Knowledge and Pleasure (London, 1747–1803)

Varenius, Bernhard, *A Compleat System of General Geography: Explaining the nature and properties of the earth* (2 vols, London, 1736)

Vaumorière, M. de, *The art of pleasing in conversation. French and English* (2 vols, London, 1735–36), ii

Waddington, Robert, *A supplement to the treatise for finding the longitude. Containing all the requisites for the ready computing and obtaining the longitude of a ship or place, by observations of the sun and moon* (London, 1764)

—— *A practical method for finding the longitude and latitude of a ship at sea, by observations of the moon; with general rules for computing the same, illustrated by examples. Together with all the necessary Tables, and their Explanations* (London, 1763)

Walker, Adam, *Syllabus of a course of lectures on natural and experimental philosophy, Viz. Astronomy, Use of the Globes, Pneumatics, Electricity, Magnetism, Chemistry, Mechanics, Hydrostatics, Hydraulics, Engineering, Fortification, Optics, &c.* (York, 1772)

Walpole, Horace, *The Yale Edition of Horace Walpole's Correspondence* (ed. W. S. Lewis) (48 vols, New Haven, 1937–83), http://www.library.yale.edu/walpole/collections/digital_collection_corr.html

Ward, John, *A Practical Method, To Discover the Longitude at Sea. By a New Contrived Automaton. Freed from all the Various Effects of Air in different Climates, &c. And not Liable to Disorder by the Irregular Motion of a Ship. The whole Method Rendered Plain and Easie to be Understood by every Mariner* (London, 1714)

Ward, William, *The description and use of a new astronomical instrument, for taking altitudes of the sun and stars at sea, without an horizon; together with an easy and sure method of observing the eclipses of Jupiter's satellites, or any other phoenomenon of the like kind, on ship-board; In order to determine the Difference of Meridians at Sea* (London, 1735)

The Weekly Journal, or British Gazetteer. Being the freshest advices foreign and domestic (London, 1715–30)

West, William, *Mathematics. With eleven copper-plates. By the late Rev. Mr. Wm. West of Exeter. Revised by John Rowe* (London, 1762)

Whiston, William, *The Longitude Discovered by the eclipses, occultations, and conjunctions of Jupiter's planets. With descriptions of those refracting and reflecting telescopes; and of those Sectors, and that Quadrant, which are the instruments necessary for this discovery* (London, 1741)

—— *The Longitude discovered by the eclipses, occultations and conjunctions of Jupiter's planets. With descriptions of those Refracting and Reflecting Telescopes; and of those Sectors, and that Quadrant, which are the Instruments necessary for this Discovery, both at Land and Sea* (London, 1738)

—— *To the Honourable the Commissioners of the Admiralty. The Memorial of William Whiston* (London, 1737)

—— *The calculation of solar eclipses without parallaxes. With a specimen of the same in the total eclipse of the sun, May 11. 1724. Now first made Publick. To which is added, A Proposal how, with the Latitude given, the Geographical Longitude of all the Parts of the Earth may be settled by the bare Knowledge of the Duration of Solar Eclipses, and especially of Total Darkness. With An Account of some late Observations made with Dipping Needles, in order to discover the Longitude and Latitude at Sea* (London, 1724)

—— *The longitude and latitude found by the inclinatory or dipping needle; wherein the laws of magnetism are also disover'd. To which is prefix'd, An Historical Preface* (London, 1721)

—— *The longitude and latitude found by the inclinatory or dipping needle* (London, 1719)

—— *A new method for discovering the longitude both at sea and land, humbly proposed to the consideration of the publick* (London, 1714)

—— *Reasons for a Bill proposing a Reward for the Discovery of the Longitude* (London, 1714)

White, Benjamin, *A catalogue of the library of The late Learned and Reverend Gloucester Ridley, D.D. Of Poplar, in Middlesex* (London, 1775)

Williams, Zachariah, *An Account of an Attempt to Discover the Longitude at Sea, by an exact theory of the Variation of the Magnetical Needle* (London, 1755)

Wilson, Henry, *Navigation new modell'd: or, a treatise of geometrical trigonometrical arithmetical instrumental and practical navigation* (London, 1723)

—— *The description and use of that most excellent invention, call'd the globular chart: Shewing its Agreeableness to the Globe, And the Natural and Easy Consequences thereof in the Practice of Navigation; with a specimen of a sea-chart in that projection; and trigonometrical calculations, to prove the Truth thereof, both in Course, Latitude, Longitude, Meridian Distance (or Departure) Distance in the Arch of a great Circle, and Distance in the Rumb, tho' so Extensive as to exceed 1200 Leagues; and all measur'd by a Scale of Equal Parts, which cannot be done upon any Projection but this only* (London, 1722)

Winstanley, William, *Poor Robin, 1665. An almanack after a new fashion being the first after bissextile, or leap-year: wherein the reader may see (if he be not blind) many remarkable things worthy of observation* (London, 1665)

Woolston, Thomas, *A second letter to the Reverend Dr. Bennet, rector of St. Giles's Cripplegate. In defence of the apostles and primitive fathers of the Church, for their allegorical interpretation of the Law of Moses, against the ministers of the letter, and literal commentators of this age* (London, 1721)

Wright, Robert, *New and correct tables of the lunar motions, according to the Newtonian theory: as it is truly freed from all errors of the press* (Manchester, 1732)

—— *An Humble Address to the Right Honourable the Lords, and the rest of the Honourable commissioners, appointed by Act of Parliament to judge of all performances relating to the longitude* (London, 1728)
Young, Edward, *Conjectures on Original Composition. In a letter to the author of Sir Charles Grandison* (1759)

Printed Secondary Works

Agnew, Jean-Christophe, *Worlds Apart: The Market and the Theatre in Anglo-American Thought, 1550–1750* (Cambridge, 1986)
Aked, Charles K., 'A Short Epistle to the Longitudinarians' in *Antiquarian Horology* 19:3 (1991), pp.292–300
Alpers, Svetlana, *The Art of Describing: Dutch Art in the Seventeenth Century* (London, 1983)
Altick, Richard D., *The Shows of London* (London, 1978)
Amelang, James, *The Flight of Icarus: Artisan Autobiography in Early Modern Europe* (Stanford, 1998)
Anderson, Julie, Barnes, Emm, and Shackleton, Emma, *The Art of Medicine: Over 2,000 Years of Images and Imagination* (Chicago, 2011)
Andrewes, William J. H., 'Bibliography of Published and Manuscript Sources relating to John Harrison' in *Antiquarian Horology* 29:4 (2006), pp.530–7
—— (ed.), *The Quest for Longitude: The Proceedings of the Longitude Symposium, Harvard University, Cambridge, Massachusetts, November 4–6 1993* (Cambridge, Massachusetts, 1996)
Andrews, Jonathan, Briggs, Asa, Porter, Roy, Tucker, Penny, and Waddington, Keir, *The History of Bethlem, 1247–1993* (London, 1997)
Armintor, Deborah Needleman, 'The Sexual Politics of Brobdingnag' in *Studies in English Literature* 47:3 (2007), pp.619–40
Ash, Eric H., 'Introduction: Expertise and the Early Modern State' in *Osiris 25:1, Expertise and the Early Modern State* (2010), pp.1–24
Baigrie, Brian S., *Picturing Knowledge: Historical and Philosophical Problems Concerning the Use of Art in Science* (Toronto, c.1996)
Baillie, G. H., *Clocks and Watches: A Historical Bibliography* (London, 1951)
Baker, Alexi, Dunn, Richard, Higgitt, Rebekah, Schaffer, Simon, and Waring, Sophie, *The Board of Longitude, 1714–1828* (forthcoming)
Balakier, Ann Stewart, and Balakier, James J., *The Spatial Infinite at Greenwich in Works by Christopher Wren, James Thornhill, and James Thomson: The Newton Connection* (Lewiston, c.1995)
Barrett, Katy, '"Explaining" Themselves: The Barrington Papers, the Board of Longitude, and the Fate of John Harrison' in *Notes and Records of the Royal Society* 65 (2011), pp.145–62
Barrett, Katy, et al., 'Looking for "the Longitude"' in *British Art Studies* 2 (Spring 2016), https://www.britishartstudies.ac.uk/issues/issue-index/issue-2/longitude
Bayly, C. A., *Empire and Information: Intelligence Gathering and Social Communication in India, 1780–1870* (Cambridge, 1996)

Beattie, Lester M., *John Arbuthnot: Mathematician and Satirist* (Cambridge, 1935)

Bender, John, *Imagining the Penitentiary: Fiction and the Architecture of Mind in Eighteenth-Century England* (Chicago, 1987)

Bender, John, and Marrinan, Michael, *The Culture of Diagram* (Stanford, 2010)

Benjamin, Marina (ed.), *A Question of Identity: Women, Science, and Literature* (New Brunswick, c.1993)

Bennett, Jim, 'Catadioptrics and Commerce in Eighteenth-Century London' in *History of Science* 44 (2006), pp.247–78

—— 'Shopping for Instruments in Paris and London' in Pamela H. Smith and Paula Findlen (eds.), *Merchants and Marvels: Commerce, Science, and Art in Early Modern Europe* (New York, 2002), pp.370–95

—— 'The Travels and Trials of Mr Harrison's Timekeeper' in Marie-Noëlle Bouguet, Christian Licoppe, and H. Otto Sibum (eds.), *Instruments, Travel and Science: Itineraries of Precision from the Seventeenth to the Twentieth Century* (London, 2002), pp.75–95

—— 'Horology, History and Harrison' in *Antiquarian Horology* (Autumn 1997), pp.451–6 [and responses: (Winter 1997), pp.558–61, (Spring 1998), pp.77–82, (Summer 1998), p.169, (Autumn 1998), p.269, (Winter 1998), p.380]

—— 'Science Lost and Longitude Found: The Tercentenary of John Harrison' in *Journal of the History of Astronomy XXIV* (1993), pp.281–7

—— 'The Longitude and the New Science' in *Vistas in Astronomy* 28:1 (1985), pp.219–25

Bennett, Jim, Cooper, Michael, Hunter, Michael, and Jardine, Lisa, *London's Leonardo: The Life and Work of Robert Hooke* (Oxford, 2003)

Berg, Maxine, and Eger, Elizabeth (eds.), *Luxury in the Eighteenth Century: Debates, Desires and Delectable Goods* (Basingstoke, 2002)

Bermingham, Ann, and Brewer, John (eds.), *The Consumption of Culture 1600–1800: Image, Object, Text* (London, 1995)

Berry, Helen, 'Rethinking Politeness in Eighteenth-Century England: Moll King's Coffee House and the Significance of "Flash Talk"' in *Transactions of the Royal Historical Society, 6th series* 11 (2001), pp.65–81

Betts, Jonathan, *Marine Chronometers at Greenwich* (Oxford, 2017)

Betts, Jonathan, and King, Andrew, 'Jeremy Thacker: Longitude Imposter?' in *Times Literary Supplement Letters* (18 March 2009)

Biagioli, Mario, 'From Print to Patents: Living on Instruments in Early Modern Europe' in *History of Science* 44 (2006), pp.139–86

—— 'Patent Republic: Representing Inventions, Constructing Rights and Authors' in *Social Research* 73:4 (2006), pp.1,129–72

Bindman, David (ed.), *Hogarth and his Times* (London, 1997)

Bindman, David, Ogée, Frédéric, and Wagner, Peter (eds.), *Hogarth: Representing Nature's Machines* (Manchester, 2001)

Bindman, David, and Wilcox, Scott (eds.), *'Among the Whores and Thieves': William Hogarth and the Beggar's Opera* (New Haven, 1997)

Black, Jeremy, *Maps and History: Constructing Images of the Past* (London, 1997)

Bogel, Fredric V., *The Difference Satire Makes: Rhetoric and Reading from Jonson to Byron* (London, 2001)
Boucé, Paul-Gabriel (ed.), *Sexuality in Eighteenth-Century Britain* (Manchester, c.1982)
Bouvier, Joseph R., *English Books on Finding Longitude at Sea to 1775: An Annotated Bibliography* (Virginia, 2000)
Brantlinger, Patrick, *Fictions of State: Culture and Credit in Britain, 1694–1994* (Ithaca, 1996)
Brewer, John, *The Pleasures of the Imagination: English Culture in the Eighteenth Century* (London, 1997)
Broman, Thomas, 'The Habermasian Public Sphere and "Science" in the Enlightenment' in *History of Science* 36 (1998), pp.301–36
Bryden, David, 'Figurative Instruments in the Georgian Era' in *Bulletin of the Scientific Instrument Society* No. 147 (2020), pp.2–10
Burke, Peter, *A Social History of Knowledge: From Gutenberg to Diderot* (Cambridge, 2000)
Bynum, William, Porter, Roy, and Shepherd, Michael (eds.), *The Anatomy of Madness: Essays in the History of Psychiatry* (3 vols, London & New York 1985)
Byrd, Max, *Visits to Bedlam: Madness and Literature in the Eighteenth Century* (Columbia, 1974)
Cantor, Geoffrey, 'Berkeley's *The Analyst* Revisited' in *Isis* 75:4 (1984), pp.668–83
Carswell, John, *The South Sea Bubble* (London, 1960)
Castle, Terry, *The Female Thermometer: Eighteenth-Century Culture and the Invention of the Uncanny* (Oxford, 1995)
—— 'Eros and Liberty at the English Masquerade, 1710–90' in *Eighteenth-Century Studies* 17:2 (1983–84), pp.156–76
Chapman, Allan, *Dividing the Circle: The Development of Critical Angular Measurement in Astronomy 1500–1850* (Chichester, 1995)
Clark, William, Golinski, Jan, and Schaffer, Simon (eds.), *The Sciences in Enlightened Europe* (Chicago, 1999)
Cook, Alan, *Edmond Halley: Charting the Heavens and the Seas* (Oxford, 1998)
Cooter, Roger, and Pumfrey, Stephen, 'Separate Spheres and Public Places: Reflections on the History of Science Popularisation and Science in Popular Culture' in *History of Science 32* (1994), pp.237–67
Crary, Jonathan, *Techniques of the Observer: On Vision and Modernity in the Nineteenth Century* (Cambridge, 1990)
Croarken, Mary, 'Tabulating the Heavens: Computing the Nautical Almanac in 18th-Century England' in *IEEE Annals of the History of Computing* 25:3 (2003), pp.48–61
Cummings, A. J. G., and Stewart, Larry, 'The Case of the Eighteenth-Century Projector: Entrepreneurs, Engineers, and Legitimacy at the Hanoverian Court in Britain' in Bruce T. Moran (ed.), *Patronage and Institutions: Science, Technology, and Medicine at the European Court 1500–1750* (Woodbridge, 1991), pp.235–61

Cunningham, Andrew, 'Sydenham versus Newton: The Edinburgh Fever Dispute in the 1690s between Andrew Browne and Archibald Pitcairne' in *Medical History. Supplement* (1981), pp.71–98

Cunningham, S., 'Bedlam and Parnassus: Eighteenth-Century Reflections' in B. Harris (ed.), *Eighteenth-Century Studies* 24 (1971), pp.36–55

Dabhoiwala, Faramerz, 'The Appropriation of Hogarth's Progresses' in *Huntington Library Quarterly* 75:4 (2012), pp.1–19

—— *The Origins of Sex: A History of the First Sexual Revolution* (London, 2012)

Dabydeen, David, *Hogarth, Walpole and Commercial Britain* (London, 1987)

—— *Hogarth's Blacks: Images of Blacks in Eighteenth Century English Art* (Manchester, 1987)

Daston, Lorraine, and Galison, Peter, *Objectivity* (London, 2007)

Davey, James, and Johns, Richard, *Broadsides: Caricature and the Navy 1756–1815* (Barnsley, 2012)

Davie, Donald, *The Language of Science and the Language of Literature 1700–1740* (London, 1963)

Day, Joan, 'The Costers: Copper-Smelters and Manufacturers' in *Transactions of the Newcomen Society* 47 (1975), pp.47–58

De Bolla, Peter, *The Education of the Eye: Painting, Landscape, and Architecture in Eighteenth-Century Britain* (Stanford, 2003)

Delbourgo, James, *A Most Amazing Scene of Wonders: Electricity and Enlightenment in Early America* (Boston, 2006)

DePorte, Michael, *Nightmares and Hobbyhorses: Swift, Sterne and Augustan Ideas of Madness* (California, 1974)

De Voogd, P. J., *Henry Fielding and William Hogarth: The Correspondences of the Arts* (Amsterdam, 1981)

Dooley, Brendan, *The Social History of Skepticism: Experience and Doubt in Early Modern Culture* (Baltimore, c.1999)

Driver, Felix, and Martins, Luciana, 'Visual Histories: John Septimus Roe and the Art of Navigation, c.1815–1830' in *History Workshop Journal* 54 (2002), pp.144–61

Dunn, Richard, 'Taming the Telescope: Notes on the Instrumental Iconography of Georgian Astronomy' in *Acta Historica Astronomiae Vol.45 Seeing Further: Essays on the Early History of the Telescope and the History of Science in Augsburg in Memory of Inge Keil* (2012), pp.191–201

—— *The Telescope: A Short History* (London, 2009)

Dunn, Richard, and Higgitt, Rebecca, *Finding Longitude* (Collins and Royal Museums Greenwich, 2014)

Edgerton Jr, Samuel Y., *The Heritage of Giotto's Geometry: Art and Science on the Eve of the Scientific Revolution* (Ithaca, 1991)

Eisenstein, Elizabeth, *The Printing Revolution in Early Modern Europe* (Cambridge, 1983)

Erwin, Timothy, 'William Hogarth and the Aesthetics of Nationalism' in *Huntington Library Quarterly* 64:3/4 (2001), pp.383–410

Fara, Patricia, 'Elizabeth Tollet: A New Newtonian Woman' in *History of Science* 40 (2002), pp.169–87

—— *Newton: The Making of Genius* (London, 2002)
—— *Sympathetic Attractions: Magnetic Practices, Beliefs, and Symbolism in Eighteenth-Century England* (Chichester, 1996)
—— '"A Treasure of Hidden Vertues": The Attraction of Magnetic Marketing' in *British Journal for the History of Science* 28:1 (1995), pp.5–35
Farrell, Maureen, 'William Whiston: The Longitude Man' in *Vistas in Astronomy* 20:1 (1976), pp.131–4
Feather, John, 'The Power of Print: Word and Image in Eighteenth-Century England' in Jeremy Black (ed.), *Culture and Society in Britain 1660–1800* (Manchester, 1997), pp.51–68
—— 'The Publishers and the Pirates: British Copyright Law in Theory and Practice, 1710–1775' in *Publishing History* 22 (1987), pp.5–32
Forbes, Eric G., *Greenwich Observatory: Vol. 1, Origins and Early History (1675–1835)* (London, 1975)
—— 'Index of the Board of Longitude Papers at the Royal Greenwich Observatory, Parts I-III' in *Journal of the History of Astronomy, Volumes 1–2* (1970–71)
Fort, Bernadette, and Rosenthal, Angela (eds.), *The Other Hogarth: Aesthetics of Difference* (Oxford, 2001)
Foucault, Michel, *Madness and Civilisation: A History of Insanity in the Age of Reason* (transl. Richard Howard) (London, 1967)
Fox, Celina, *The Arts of Industry in the Age of Enlightenment* (New Haven, c.2009)
Freiburg, Rudolf, 'Our never failing guide, the Watch: Der "Längengrad" in der englischen Literatur und Kultur des achzehntehn Jarhunderts' in Rudolf Freiburg, Christian Lubkoll, and Harald Neymeyer (eds.), *Zwischen Literatur und Naturwissenschaft* (Berlin, 2017), pp.33–62
Furbank, P. N., 'Misreading Gulliver's Travels: Is Jonathan Swift's famous satire a defence of humanity rather than a condemnation of it?' in *Times Literary Supplement* (10 November 2010)
Fussell, Paul, *The Rhetorical World of Augustan Humanism: Ethics and Imagery from Swift to Burke* (Oxford, 1965)
Gatrell, Vic, *City of Laughter: Sex and Satire in Eighteenth-Century London* (London, 2006)
George, M. D., *Hogarth to Cruikshank: Social Change in Graphic Satire* (London, 1967)
Gerrard, Ted, *Astronomical Minds: The True Longitude Story* (Broadford, 2007)
Goldgar, Anne, *Impolite Learning: Conduct and Community in the Republic of Letters, 1680–1750* (Yale, 1995)
Goldgar, B. A., *Walpole and the Wits: The Relation of Politics to Literature 1722–1742* (Lincoln, 1976)
—— *The Curse of Party: Swift's Relations with Addison and Steele* (Lincoln, 1961)
Gould, Rupert, *John Harrison and his Timekeepers* (London, 1958)
—— *The Marine Chronometer: Its History and Development* (London, 1923)
Gowing, Lawrence, *Hogarth* (London, 1971)
Griffin, Amy, Macaro, Gabriella, and Townsend, Joyce, *The Development of Hogarth's Series* A Rake's Progress*: From Paintings to Prints*, https://www.tate.org.uk/research/features/hogarth-rakes-progress-paintings-to-prints

Guerrini, Anita, 'Anatomists and Entrepreneurs in Early Eighteenth-Century London' in *Journal of the History of Medicine and Allied Sciences* 59:2 (2004), pp.219–39

—— 'The Tory Newtonians: Gregory, Pitcairne and their Circle' in *Journal for British Studies* 25:3 (1986), pp.288–311

Guest, Harriet, *A Form of Sound Words: The Religious Poetry of Christopher Smart* (Oxford, c.1989)

Habermas, Jürgen, *The Structural Transformation of the Public Sphere: An Inquiry into a Category of Bourgeois Society* (transl. Thomas Burger and Frederick Lawrence) (Cambridge, MA, 1991)

Hacking, Ian, *Mad Travellers: Reflections on the Reality of Transient Mental Illnesses* (London, 1998)

Hall, Marie Boas, *Promoting Experimental Learning: Experiment and the Royal Society 1660–1727* (Cambridge, 1991)

Hallett, Mark, *Hogarth* (London, 2000)

—— *The Spectacle of Difference: Graphic Satire in the Age of Hogarth* (London, 1999)

—— 'The Medley Print in Early Eighteenth-Century London' in *Art History* 20.2 (1997), pp.214–37

Hanson, Craig Ashley, *The English Virtuoso: Art, Medicine, and Antiquarianism in the Age of Empiricism* (Chicago, c.2009)

Hawes, Clement, *Mania and Literary Style: The Rhetoric of Enthusiasm from the Ranters to Christopher Smart* (Cambridge, 2005)

Haynes, Rosalind D., *From Faust to Strangelove: Representations of the Scientist in Western Literature* (London, 1994)

Hayton, David, 'Sir Richard Cocks: The Political Anatomy of a Country Whig' in *Albion: A Quarterly Journal Concerned with British Studies* 20:2 (1988), pp.221–46

Hewish, John, *Rooms near Chancery Lane: The Patent Office under the Commissioners, 1852–1883* (London, c.2000)

Heyd, Michael, *'Be sober and reasonable': The Critique of Enthusiasm in the Seventeenth and Early Eighteenth Centuries* (Leiden, 1995)

Historien af Mr. Harrisons Forsøg til Længdens Opfindelse formedelst et Uhr eller en Tidmaaler, samt hvad der ved dets Prøving og Bedømmelse i Engeland i vore Tider er foregaaet, saavelsom og en Oversættelse af Principperne tilligemed Tegningerne af samme Tidmaaler, udgivne i Aaret 1767 efter Ordre af Commisarierne for Længden (Kiøbenhavn 1768. Trykt hos Nicolaus Møller, Kgl. Hof-Bogtrykker)

Hoppitt, Julian, 'The Myths of the South Sea Bubble' in *Transactions of the Royal Historical Society, 6th series, 12* (2002), pp.141–65

Howse, Derek, 'Britain's Board of Longitude: The Finances, 1714–1828' in *Mariner's Mirror* 84:4 (1998), pp.400–17

—— *Greenwich Time and the Longitude* (London, 1997)

Hume, Robert D., 'The Economics of Culture in London, 1660–1740' in *Huntington Library Quarterly* 69:4 (2006), pp.487–533

Hunter, David, 'Copyright Protection for Engravings and Maps in Eighteenth-Century Britain', *Library* 6 (1987), pp.128–47

Ingold, Tim, *Lines: A Brief History* (Abingdon, 2007)

Ivins, William M., *Prints and Visual Communication* (Cambridge, 1969)

Jacob, Christian, *The Sovereign Map: Theoretical Approaches in Cartography throughout History* (transl. Tom Conley, ed. Edward H. Dahl) (London, 2006)

Jacob, Margaret, *The Newtonians and the English Revolution 1689–1720* (Ithaca, 1976)

Jacob, Margaret, and Stewart, Larry (eds.), *Practical Matter: Newton's Science in the Service of Industry and Empire 1687–1851* (Cambridge, 2004)

Johns, Adrian, *Piracy: The Intellectual Property Wars from Gutenberg to Gates* (Chicago, 2009)

Johnson, Peter, 'The Board of Longitude 1714–1828' in *Journal of the British Astronomical Association* 99:2 (1989), pp.63–9

Jones, C., and Galison, P. (eds.), *Picturing Science, Producing Art* (London, 1998)

Jones, Richard Foster, *Ancients and Moderns: A Study of the Background of the Battle of the Books* (St. Louis, 1936)

Jordanova, Ludmilla, *The Look of the Past: Visual and Material Evidence in Historical Practice* (Cambridge, 2012)

—— *Sexual Visions: Images of Gender in Science and Medicine between the Eighteenth and Twentieth Centuries* (New York, 1989)

Kernan, Alvin, *Printing Technology, Letters and Samuel Johnson* (Princeton, 1987)

King, Andrew, 'The Images of John Harrison' in *Antiquarian Horology* (June 2006), pp.538–43

Köberer, Wolfgang, 'German Contributions to Solving the Longitude Problem in the Seventeenth and Eighteenth Centuries' in *The Mariner's Mirror* 108:3, pp.262–85

Koon, Helene, 'Eliza Haywood and the "Female Spectator"' in *Huntington Library Quarterly* 42:1 (Winter, 1978), pp.43–55

Kromm, Jane, *The Art of Frenzy: Public Madness in the Visual Culture of Europe, 1500–1850* (London, 2002)

Krysmanski, Bernd, *Hogarth's Hidden Parts: Satiric Allusion, Erotic Wit, Blasphemous Bawdiness and Dark Humour in Eighteenth-Century English Art* (Hildesheim, 2010)

—— 'Hogarth's *A Rake's Progress*: An "Anti-Passion" in Disguise' in *1650–1850: Ideas, Aesthetics, and Inquiries in the Early Modern Era 4* (1998), pp.137–82

Kuhn, Albert, 'Dr Johnson, Zachariah Williams, and the Eighteenth-Century Search for the Longitude' in *Modern Philology* 82 (1984), pp.40–52

Landow, George P., *Images of Crisis: Literary Iconology, 1750 to the Present* (Boston, 1982)

Latour, Bruno, *Pandora's Hope: Essays on the Reality of Science Studies* (London, 1999)

—— *Science in Action: How to Follow Scientists and Engineers through Society* (Milton Keynes, 1987)

—— 'Visualisation and Cognition: Drawing Things Together', www.bruno-latour.fr No.21

Law, John, 'On the Methods of Long Distance Control: Vessels, Navigation and the Portuguese Route to India' in John Law (ed.), *Power, Action and Belief: A New Sociology of Knowledge? Sociological Review Monograph* 32 (Henley, 1986), pp.234–63

Lefèvre, Wolfgang, *Picturing Machines 1400–1700* (Massachusetts, 2004)

Legg, Stephen, and Brown, Michael, 'Moral Regulation: Historical Geography and Scale' in *Journal of Historical Geography* (2013), In Press, http://dx.doi.org/10.1016/j.jhg.2013.02.001

Levine, Joseph M., *Between the Ancients and the Moderns: Baroque Culture in Restoration England* (Yale, 1999)

—— *The Battle of the Books: History and Literature in the Augustan Age* (New York, 1991)

—— *Dr Woodward's Shield: History, Science and Satire in Augustan England* (Berkeley, 1977)

Livingstone, David N., *Putting Science in its Place: Geographies of Scientific Knowledge* (Chicago, 2003)

—— *The Geographical Tradition: Episodes in the History of a Contested Enterprise* (Oxford, 1992)

Loftis, John, 'Richard Steele's Censorium' in *Huntingdon Library Quarterly* 14 (1950–51), pp.43–66

Lucas, Anya, Johns, Richard, Stewart, Sophie, and Paine, Stephen, *The Painted Hall: Sir James Thornhill's Masterpiece at Greenwich* (London, 2019)

Lynall, Gregory, 'Scriblerian Projections of Longitude: Arbuthnot, Swift, and the Agency of Satire in a Culture of Invention' in *Journal of Literature and Science* Volume 7, No. 2 (2014), pp.1–18

—— *Swift and Science: The Satire, Politics, and Theology of Natural Knowledge, 1690–1730* (New York, 2012)

Macey, Samuel L., 'Hogarth and the Iconography of Time' in *Studies in Eighteenth-Century Culture* 5 (1976), pp.41–53

MacGregor, John M., *The Discovery of the Art of the Insane* (Princeton, 1989)

Macleod, Christine, *Inventing the Industrial Revolution: The English Patent System, 1660–1800* (Cambridge, 1988)

Mandelbrote, Scott, 'The Heterodox Career of Fatio de Duillier' in John Brooke and Ian Maclean (eds.), *Heterodoxy in Early Modern Science and Religion* (2005), pp.263–96

McDermott, Anne (ed.), *Ashgate Critical Essays on Early English Lexicographers, Volume 5, The Eighteenth Century* (Farnham, 2012)

McIntosh, Carey, 'Eighteenth-Century English Dictionaries and the Enlightenment' in *The Yearbook of English Studies* 28 (1998), pp.3–18

McKendrick, Neil, Brewer, John, and Plumb, J. H., *The Birth of a Consumer Society: The Commercialization of Eighteenth-Century England* (London, 1982)

Midelfort, H. C. Erik, 'Reading and Believing: On the Reappraisal of Michel Foucault' in Arthur Still and Irving Velody (eds.), *Rewriting the History of Madness: Studies in Foucault's* Histoire de la folie (London, 1992), pp.105–9

Millburn, John R., *Benjamin Martin: Author, Instrument-Maker, and Country Showman* (Leyden, 1976)

Miller, Elizabeth, *Hand-Coloured British Prints* (London, 1958)
Mitchell, Elizabeth Kathleen, *Death by Hogarth* (Cambridge, MA, c.1999)
Moreland, Carl, and Bannister, David, *Antique Maps* (Oxford, 1989)
Morrison-Low, A. D., *Making Scientific Instruments in the Industrial Revolution* (Aldershot, 2007)
Morton, A. Q., 'Lectures on Natural Philosophy in London, 1750–1765: S. C. T. Demainbray (1710–1782) and the "Inattention" of His Countrymen' in *British Journal for the History of Science* 23:4 (1990), pp.411–34
Morzer Bruyns, W. F. J., *Sextants at Greenwich: A catalogue of the mariner's quadrants, mariner's astrolabes, cross-staffs, backstaffs, octants, sextants, quintants, reflecting circles and artificial horizons in the National Maritime Museum, Greenwich* (Oxford, 2009)
Mount, Harry, 'Morality, Microscopy and the Moderns: The Meaning of Minuteness in Shaftesbury's Theory of Painting' in *British Journal for Eighteenth-Century Studies* 21:2 (1998), pp.125–41
Mueller-Sievers, Helmüt, 'Tidings of the Earth: Towards a History of Romantic *Erdkunde*' in Martha B. Helfer (ed.), *Rereading Romanticism* (Amsterdam, 2000), pp.47–73
Myer, Valerie Grosvenor (ed.), *Laurence Sterne: Riddles and Mysteries* (London, c.1984)
Myrone, Martin, *Bodybuilding: Reforming Masculinities in British Art 1750–1810* (New Haven, 2005)
Nicholson, Colin, *Writing and the Rise of Finance: Capital Satires of the Early Eighteenth Century* (Cambridge, 1994)
Nicolson, Marjorie, *Science and Imagination* (Ithaca, 1956)
Nicolson, Marjorie, and Rousseau, G. S., *'This Long Disease My Life': Alexander Pope and the Sciences* (Princeton, 1968)
O'Connell, Sheila, *London 1753* (London, 2003)
Ogborn, Miles, *Spaces of Modernity: London's Geographies 1680–1780* (London, 1998)
Ogborn, M., and Withers, C. W. J. (eds.), *Georgian Geographies: Essays on Space, Place and Landscape in the Eighteenth Century* (Manchester, 2004)
Ogée, Frédéric (ed.), *The Dumb Show: Image and Society in the Works of William Hogarth* (Oxford, 1997)
Olby, R. C., Cantor, G. N., Christie, J. R. R., and Hodge, M. J. S. (eds.), *Companion to the History of Modern Science* (London, 1996)
Olson, Richard G., 'Tory High-Church Opposition to Science and Scientism in the Eighteenth Century: The Works of John Arbuthnot, Jonathan Swift, and Samuel Johnson' in John G. Burke (ed.), *The Uses of Science in the Age of Newton* (Berkeley, 1983), pp.171–204
Panofsky, Erwin, Klibansky, Raymond, and Saxl, Fritz, *Saturn and Melancholy: Studies in the History of Natural Philosophy, Religion, and Art* (London, 1964)
Paulson, Ronald, *Breaking and Remaking: Aesthetic Practice in England, 1700–1820* (London, 1989)
—— *Hogarth's Graphic Works* (London, 1989)
—— *Popular and Polite Art in the Age of Hogarth and Fielding* (Notre Dame, c.1979)

—— *Emblem and Expression: Meaning in English Art of the Eighteenth Century* (London, c.1975)
—— *Rowlandson: A New Interpretation* (London, 1972)
—— *Hogarth: His Life, Art and Times* (2 vols, New Haven, 1971)
—— *Satire and the Novel in Eighteenth-Century England* (London, 1967)
Pauwels, Luke (ed.), *Visual Cultures of Science: Rethinking Representational Practices in Knowledge Building and Scientific Communication* (Hanover, 2006)
Peakman, Julie, *Mighty Lewd Books: The Development of Pornography in Eighteenth-Century England* (Palgrave, 2003)
Phillips, Patricia, *The Scientific Lady: A Social History of Women's Scientific Interests 1520–1918* (London, c.1990)
Piper, Karen, *Cartographic Fictions: Maps, Race, and Identity* (New Brunswick, c.2002)
Pomian, Krzysztof, 'Medals/Shells=Erudition/Philosophy' in *Collectors and Curiosities: Paris and Venice, 1500–1800* (transl. Elizabeth Wiles-Portier) (Cambridge, 1990)
Poole, S., *The Politics of Regicide in England, 1750–1860: Troublesome Subjects* (Manchester, 2000)
Poovey, Mary, *A History of the Modern Fact: Problems of Knowledge in the Sciences of Wealth and Society* (Chicago, 1998)
Porter, Roy, *Bodies Politic: Disease, Death and Doctors in Britain, 1650–1900* (London, 2001)
—— 'Bethlem/Bedlam: Methods of Madness?' in *History Today* 47:10 (1997), pp.41–7
—— *Health for Sale: Quackery in England 1660–1850* (Manchester, 1989)
—— *Mind-forg'd Manacles: A History of Madness in England from the Restoration to the Regency* (London, 1987)
—— *A Social History of Madness: Stories of the Insane* (London, 1987)
—— 'Love, Sex and Madness in Eighteenth-Century England' in *Social Research* 53:2 (1986), pp.211–42
—— 'The Rage of Party: A Glorious Revolution in English Psychiatry' in *Medical History* 27 (1983), pp.35–50
Pottage, Alain, and Sherman, Brad, *Figures of Invention: A History of Modern Patent Law* (Oxford, 2010)
Pressly, William L., *James Barry's Murals at the Royal Society of Arts: Envisioning a New Public Art* (Cork, 2014)
Pumfrey, Stephen, *Latitude and the Magnetic Earth* (Cambridge, c.2002)
Quill, Humphrey, *John Harrison, Copley Medallist, and the £20,000 Longitude Prize* (Wadhurst, 1976)
—— *John Harrison: The Man who Found Longitude* (London, 1966)
Quilley, Geoff, 'The Analysis of Deceit: Sandby's Satires against Hogarth' in *Paul Sandby: Picturing Britain* (London, 2009), pp.38–47
Ratcliff, Jessica, 'Art to Cheat the Common-Weale: Inventors, Projectors, and Patentees in English Satire, ca.1630–70' in *Technology and Culture* 53:2 (2012), pp.337–65

Raven, James, Tadmore, Naomi, and Small, Helen (eds.), *The Practice and Representation of Reading in Britain 1500–1900* (Cambridge, 1996)

Raymond, Joad, *Pamphlets and Pamphleteering in Early-Modern Britain* (Cambridge, 2003)

—— *News, Newspapers and Society in Early Modern Britain* (London, 1999)

Robinson, A. H. W., *Marine Cartography in Britain: A History of the Sea Chart to 1855* (Leicester, 1962)

Rogers, Nicholas, *Whigs and Cities: Popular Politics in the Age of Walpole and Pitt* (Oxford, 1989)

Rogers, Pat, *Documenting Eighteenth-Century Satire: Pope, Swift, Gay, and Arbuthnot in Historical Context* (Newcastle, 2012)

—— 'Longitude forged: How an eighteenth-century hoax has taken in Dava Sobel and other historians' in *Times Literary Supplement* (12 November 2008)

—— *Eighteenth Century Encounters: Studies in Literature and Society in the Age of Walpole* (Brighton, 1985)

—— *Literature and Popular Culture in Eighteenth-Century England* (Brighton, 1985)

—— 'Shapes and Postures: Swift, Walpole and the Rope-Dancers' in *Literature and Popular Culture in Eighteenth-Century England* (Brighton, 1985)

—— *Grub Street: Studies in a Subculture* (London, 1972)

Rose, Gillian, *Feminism and Geography: The Limits of Geographical Knowledge* (Cambridge, 1993)

Rose, Mark, 'The Author as Proprietor: Donaldson v. Becket and the Genealogy of Modern Authorship' in *Representations* 23 (1988), pp.51–85

Rousseau, G. S., '"Wicked Whiston" and the Scriblerians: Another Ancients-Modern Controversy' in John Yolton and Leslie Ellen Brown (eds.), *Studies in Eighteenth-Century Culture* 17 (1987), pp.17–44

—— 'Science Books and their Readers in the Eighteenth Century' in Isabel Rivers (ed.), *Books and their Readers in Eighteenth-Century England* (Leicester, 1982), pp.197–255

Rousseau, G. S., and Porter, Roy, *Sexual Underworlds of the Enlightenment* (Chapel Hill, 1988)

Rusnock, Andrea, 'Correspondence Networks and the Royal Society, 1700–1750' in *British Journal for the History of Science* 32 (1999), pp.155–69

Russell, David, *Scenes from Bedlam: A History of Caring for the Mentally Disordered at Bethlem Royal Hospital and the Maudsley* (London, c.1997)

Saunders, David, *Authorship and Copyright* (London, 1992)

Schaffer, Simon, '"On Seeing Me Write": Inscription Devices in the South Seas' in *Representations* 97:1 (2007), pp.94–126

—— 'The Show That Never Ends: Perpetual Motion in the Early Eighteenth Century' in *British Journal for the History of Science* 28:2 (1995), pp.157–89

—— 'Defoe's Natural Philosophy' in John Christie and Sally Shuttleworth (eds.), *Nature Transfigured: Science and Literature, 1700–1900* (Manchester, 1989), pp.13–44

—— 'Natural Philosophy and Public Spectacle in the Eighteenth Century' in *History of Science* 21 (1983), pp.1–43

Schiebinger, Londa, *Nature's Body: Sexual Politics and the Making of Modern Science* (London, 1994)

—— *The Mind has No Sex? Women in the Origins of Modern Science* (London, 1989)

Scull, Andrew T., *Museums of Madness: The Social Organisation of Insanity in Nineteenth-Century England* (London, 1979)

Secord, James, 'How Scientific Conversation became Shop Talk' in *Transactions of the Royal Historical Society* 17 (2007), pp.129–56

Shapin, Steven, *A Social History of Truth: Civility and Science in Seventeenth-Century England* (Chicago, 1994)

Shea, William R. (ed.), *Science and the Visual Image in the Enlightenment* (Canton, 2000)

Siegel, J. R., 'Law and Longitude' in *Tulane Law Review* 84:1 (2009), pp.1–66

Smiles, Sam, *Eye Witness: Artists and Visual Documentation in Britain 1770–1830* (Aldershot, c.2000)

Smith, Pamela H., *The Body of the Artisan: Art and Experience in the Scientific Revolution* (Chicago, 2004)

Snobelen, Steven David, 'William Whiston, Isaac Newton and the Crisis of Publicity' in *Studies in History and Philosophy of Science* 35 (2004), pp.573–603

Snodin, Michael (ed.), *Rococo: Art and Design in Hogarth's England* (London, 1984)

Sobel, Dava, *Longitude: The True Story of a Lone Genius who Solved the Greatest Scientific Problem of his Time* (London, 1995)

Solkin, David H., *Painting for Money: The Visual Arts and the Public Sphere in Eighteenth-Century England* (New Haven, 1993)

Sorrenson, Richard, 'George Graham: Visible Technician' in *British Journal for the History of Science* 32 (1999), pp.203–21

—— 'The Ship as a Scientific Instrument in the Eighteenth Century' in *Osiris* (1996), pp.221–36

Soulier-Détis, Elisabeth, *'Guess at the Rest': Cracking the Hogarth Code* (Cambridge, 2010)

Spacks, Patricia Meyer, 'Ev'ry Woman is at Heart a Rake' in *Eighteenth-Century Studies* 8:1 (1974), pp.27–46

Stafford, Barbara Maria, *Artful Science: Enlightenment Entertainment and the Eclipse of Visual Education* (Cambridge, c.1994)

Stallybrass, Peter, and White, Allon, *The Politics and Poetics of Transgression* (London, 1986)

Starkie, Andrew, *The Church of England and the Bangorian Controversy 1716–1721* (Woodbridge, 2007)

Steele, Ian, *The English Atlantic 1675–1740: An Exploration of Communication and Community* (Oxford, 1986)

Steintrager, James A., 'Perfectly Inhuman: Moral Monstrosity in Eighteenth-Century Discourse' in *Eighteenth-Century Life* 21 (1997), pp.114–32

Stephens, F., and George, M. Dorothy, *Catalogue of Political and Personal Satires Preserved in the Department of Prints and Drawings in the British Museum* (11 vols, London, 1870–1954)

Stevenson, Christine, 'Hogarth's Mad King and His Audiences' in *History Workshop Journal* 49 (2000), pp.24–43

Stewart, Larry, 'Other Centres of Calculation, or, Where the Royal Society Didn't Count: Commerce, Coffee-Houses and Natural Philosophy in Early Modern London' in *British Journal for the History of Science* 32:2 (1999), pp.133–53

—— *The Rise of Public Science: Rhetoric, Technology, and Natural Philosophy in Newtonian Britain, 1660–1750* (Cambridge, 1992)

—— 'Public Lectures and Private Patronage in Newtonian England' in *Isis* 77:1 (1986), pp.47–58

—— 'The Edge of Utility: Slaves and Smallpox in the Early Eighteenth Century' in *Medical History* 29 (1985), pp.54–70

Suarez, Michael F., and Turner, Michael L. (eds.), *The Cambridge History of the Book in Britain, Volume V 1695–1830* (Cambridge, 2009)

Taylor, E. G. R., 'A Reward for the Longitude' in *Mariner's Mirror 45* (1959), pp.59–66

Thrower, Norman J. W., *The Three Voyages of Edmond Halley in the* Paramore *1698–1701* (London, 1981)

Treadwell, J. M., 'Jonathan Swift: The Satirist as Projector' in *Texas Studies in Literature and Language* 17:2 (1975), pp. 439–60

Turnbull, David, *Masons, Tricksters and Cartographers: Comparative Studies in the Sociology of Scientific and Indigenous Knowledge* (Amsterdam, 2000)

Turner, A. J., 'Berthoud in England, Harrison in France: The Transmission of Horological Knowledge in 18th-Century Europe' in *Antiquarian Horology* (Autumn, 1992), pp.219–38

Tyacke, Sarah, *London Map-sellers, 1660–1720: A collection of advertisements for maps placed in the 'London Gazette' 1668–1719, with biographical notes on the map-sellers* (Tring, 1978)

Uglow, Jenny, *Hogarth: A Life and a World* (London, 1997)

Vermeulen, Ingrid R., *Picturing Art History: The Rise of the Illustrated History of Art in the Eighteenth Century* (Amsterdam, 2011)

Viseltear, Arthur J., 'Joanna Stephens and the Eighteenth Century Lithontriptics: A Misplaced Chapter in the History of Therapeutics' in *Bulletin of the History of Medicine* 42:3 (1968), pp.199–220

Wagner, Peter, *Reading Iconotexts: From Swift to the French Revolution* (London, 1995)

—— 'Eroticism in Graphic Art: The Case of William Hogarth', *Studies in Eighteenth-Century Culture* 21 (1992), pp.53–75

—— 'Hogarth's Graphic Palimpsests: Intermedial Adaptation of Popular Literature' in *Word and Image* 7:4 (1991), pp.329–47

—— *Eros Revived: Erotica of the Enlightenment in England and America* (London, 1988)

—— 'The Veil of Medicine and Morality: Some Pornographic Aspects of the *Onania*' in *British Journal for Eighteenth-Century Studies 6* (1983), pp.179–84

Wahrman, Dror, *Mr Collier's Letter Racks: A Tale of Art and Illusion at the Threshold of the Modern Information Age* (Oxford, 2012)

Walters, Alice Nell, 'Ephemeral Events: English Broadsides of Early Eighteenth-Century Solar Eclipses' in *History of Science* XXXVII (1999), pp.1–43

—— 'Conversation Pieces: Science and Politeness in Eighteenth-Century England' in *History of Science* 35 (1997), pp.121–54

Wardhaugh, Benjamin, *Poor Robin's Prophecies: A Curious Almanac, and the Everyday Mathematics of Georgian England* (Oxford, 2012)

Wardroper, John, *The Caricatures of George Cruikshank* (London, 1977)

Weinbrot, Howard D., *The Formal Strain: Studies in Augustan Imitation and Satire* (Chicago, 1969)

Werrett, Simon, *Fireworks: Pyrotechnic Arts and Sciences in European History* (Chicago, 2010)

Wigelsworth, Jeffrey R., *Selling Science in the Age of Newton: Advertising and the Commoditization of Knowledge* (Abingdon, 2010)

—— 'Bipartisan Politics and Practical Knowledge: Advertising of Public Science in Two London Newspapers, 1695–1720' in *British Journal for the History of Science* 41 (2008), pp.517–40

—— 'Navigation and Newsprint: Advertising Longitude Schemes in the Public Sphere ca.1715' in *Science in Context* 21:3 (2008), pp.351–76

Williams, Glyn, *The Prize of All the Oceans: The Triumph and Tragedy of Anson's Voyage Round the World* (London, 1999)

Williams, Raymond, *Keywords: A Vocabulary of Culture and Society* (London, 1983)

Wimsatt, W. K., *Philosophic Words: A Study of Style and Meaning in the Rambler and Dictionary of Samuel Johnson* (Hamden, 1968)

Wintroub, Michael, 'The Heavens Inscribed: The Instrumental Poetry of the Virgin in Early Modern France' in *British Journal for the History of Science* 42:2 (2009), pp.161–85

Wittkower, Rudolph, and Wittkower, Margot, *Born under Saturn: The Character and Conduct of Artists: A Documented History from Antiquity to the French Revolution* (London, 1963)

Womersley, David (ed.), *A Companion to Literature from Milton to Blake* (Oxford, 2000)

—— (ed.), *Augustan Critical Writing* (London, 1997)

Womersley, David, Bullard, Paddy, and Williams, Abigail (eds.), *'Cultures of Whiggism': New Essays on English Literature and Culture in the Long Eighteenth Century* (Newark, 2005)

Woodcroft, Bennet, *Alphabetical Index of Patentees of Inventions from March 2, 1617 to October 1, 1852* (London, 1969)

Woodward, David, *Art and Cartography: Six Historical Essays* (Chicago, 1987)

Yamamoto, Koji, 'Piety, Profit and Public Service in the Financial Revolution' in *English Historical Review* 125.521 (2011)

Zitin, Abigail, 'Thinking Like an Artist: Hogarth, Diderot, and the Aesthetics of Technique' in *Eighteenth-Century Studies* 46:4 (2013), pp.555–70

Unpublished Theses

Barrett, Katy, 'The Wanton Line: Hogarth and the Public Life of Longitude' (Cambridge Univ. Ph.D. Thesis, 2014)

Grant, Florence E. H., 'The Philosophical Instruments of George III: A Case Study in the Material Culture of Experimental Philosophy' (King's College London Ph.D. Thesis, 2011)

Johns, Richard, 'James Thornhill and Decorative History Painting in England after 1688' (University of York Ph.D. Thesis, 2004)

Orrje, Jacob, 'Navigating the Neighbourhoods of Longitude: Bengt Ferrner's (1724–1802) Journal of London's Astronomical Community 1759–60' (Cambridge Univ. M.Phil. Thesis, 2011)

Phillips, Eóin, 'Making Time Fit: Astronomers, Artisans and the State, 1770–1820' (Cambridge Univ. Ph.D. Thesis, 2014)

Snobelen, S. D., 'William Whiston: Natural Philosopher, Prophet, Primitive Christian' (Cambridge Univ. Ph.D. Thesis, 2000)

Walters, Alice Nell, 'Tools of Enlightenment: The Material Culture of Science in Eighteenth-Century England' (University of California, Berkeley Ph.D. Thesis, 1992)

Waring, Sophie, 'Thomas Young, the Board of Longitude and the Age of Reform' (Cambridge Univ. Ph.D. Thesis, 2015)

Index

www.ingramcontent.com/pod-product-compliance
Lightning Source LLC
LaVergne TN
LVHW011817200926
848484LV00005B/1221
* 9 7 8 1 8 0 2 0 7 0 5 3 8 *